THE BRITISH IS

THEMES AND CASE STUDIES

CHRIS SMART

HODDER AND STOUGHTON
LONDON SYDNEY AUCKLAND TORONTO

CONTENTS

SECTION 1 RESOURCES
- Coal — 2
- Gas and oil — 8
- Limestone — 12
- Water — 16
- Energy — 20
- Fishing — 28
- Forestry — 32

SECTION 2 FARMING
- Dairy farming — 38
- Arable farming — 42
- Hill farming — 46
- Market gardening — 48

SECTION 3 INDUSTRY
- Iron and steel — 50
- Oil refining — 56
- Chemicals — 58
- Car industry — 62
- Textiles — 66
- Shipbuilding — 70
- Light industry — 72
- Change – a North/South divide? — 76

SECTION 4 TRANSPORT
- Roads — 78
- Railways — 82
- Ports — 88
- Airports — 92

SECTION 5 POPULATION AND SETTLEMENT
- Population — 96
- Settlement — 98
- Conurbations — 102
- New Towns — 106
- Regional centres — 110
- Holiday resorts — 112
- Areas where few people live — 114
- National Parks — 118

THE ROUND BRITAIN RACE — 120

MAP READING QUESTIONS — 122

ORDNANCE SURVEY MAP EXTRACTS — 123

INDEX — 126

ISBN 0 340 26901 4

First published 1987

Copyright © 1987 Chris Smart

All rights reserved. No part of this publication may be reproduced or transmitted in any form or by any means, electronic, or mechanical, including photocopy, recording, or any information storage and retrieval system, without permission in writing from the publisher or under licence from the Copyright Licensing Agency Limited. Details of such licences (for reprographic reproduction) may be obtained from the Copyright Licensing Agency Limited, of 33–34 Alfred Place, London, WC1E 7DP.

Typeset by Rowland Phototypesetting Ltd, Bury St Edmunds, Suffolk.
Printed for Hodder and Stoughton Educational a division of Hodder and Stoughton Ltd., Mill Road, Dunton Green, Sevenoaks, Kent, by Colorcraft Ltd., Hong Kong.

The author and publishers thank the following for permission to reproduce photographs in this book:

Aerofilms (3.44, 4.44, 5.17, 5.20), Austin Rover (3.23, 3.27, 3.29, 3.31), Bath City Council/Civic Trust Library (1.31), BBC Hulton Picture Library (3.22), Bluebell Railway (4.15), Bluecrest Foods Ltd (1.68), Janet & Colin Bord (4.18, 4.19, 5.6), Brighton Tourist Services (5.36), Britain on View (BTA/ETB) (4.3, 4.16), British Petroleum (1.27), British Rail (4.20, 4.27, 4.28, 4.32), British Steel (3.3, 3.7), Robert Cartwright (1.70, 1.71, 1.73, 1.74, 2.6, 2.9, 2.19, 2.20), COI (4.6), Farmer's Weekly (2.5, 2.24), Nance Fyson (p. 1, p. 105 top), Handford Photography (1.34, 1.36, 4.35, 4.43), Highland and Islands Tourist Board (9.39, 5.41, 5.42), ICI (3.20), International Wool Secretariat (3.34, 3.35, 3.36), David Jones (1.55, 3.14, 5.46), A. Mellors (4.10, 4.12), Milk Marketing Board (2.11, 2.12), John Mills Photography (5.27, 5.28, 5.29, 5.30), Museum of Natural History (1.33), National Coal Board (1.1, 1.3, 1.4, 1.5, 1.6, 1.13, 1.14, 1.17, 1.20), National Farmers Union (2.3, 2.15, 2.16, 2.25, 2.26), Nissan (3.25), Oxford Scientific Films (2.13, 2.17, 2.18, 3.37), Pictor International (3.9), Port of Liverpool Authority (4.31), Redditch Development Corporation (3.46), M. Reyner (4.11), Shell (4.33), Chris Smart (5.7, 5.20), UK Atomic Energy Authority (1.48, 1.51, 1.52), Unilever (3.18), Derek G. Widdicombe (title page, 1.62, 1.63).

Title page photograph:
Hadrian's Wall, Northumberland

SECTION 1 · Resources

Every country has some resources. A resource is something that can be obtained from the ground, water or even air. These may be minerals, natural gas or even forests. Nations have resources in varying amounts. Some countries have far more resources than others. Even countries with rich reserves of coal may still be very short of, say, iron ore. In Britain, for example, there are 120 million tonnes of coal mined each year, but only 5200 tonnes of tin. Britain also has to import many minerals, like bauxite, because the UK has no deposits.

Importing and exporting helps many countries gain a fairer share of the world's resources. But the world population is continually growing. Some resources are being used up very quickly. Many resources are finite, meaning that one day they will run out.

Oil and natural gas are examples of finite resources. It is thought that oil and gas from the North Sea will run out shortly after the year 2000. We shall have to find other forms of energy to take their place. Britain's coal reserves should last for several hundred years at the present rate of production. Another answer may be to try and make more use of infinite resources, like water, wind and sunlight. New HEP (hydroelectric power) schemes, nuclear energy, and wind and solar power may all help to provide our future energy needs. Some new ideas, like wave machines in the sea, may seem a little far-fetched, but imagine you were living in the year 1066. What would you have thought Britain's resources were then? Would you have thought of North Sea oil?

Questions

1 What is meant by 'a resource'? Name four other examples of resources not listed on this page.

2 What is meant by the terms 'finite resources' and 'infinite resources'? Draw two columns, one for 'finite resources' and the other for 'infinite resources'. List six finite resources and four infinite resources. Why is it difficult to name more infinite resources?

3 What is the difference between resources and reserves? In Britain, tin used to be mined in Cornwall. Mining then stopped, but a couple of mines re-opened a few years ago and some of the waste tips were reworked for tin. Can you suggest why this might have happened?

4 The woman in the photo (left) is putting bottles into a Bottle Bank. The glass will be recycled and used again. What other materials are collected in Britain for recycling? How does recycling help a country's resources to go further?

Coal

○ INTRODUCTION

In Britain, at the turn of the century, 250 million tonnes of coal were mined annually. One third of this was exported. After 1900, production began to decline, with very little being sent abroad. This was mainly due to the increased availability and popularity of cleaner fuels, such as natural gas and oil. During the last ten years these have become more expensive. Future supplies are also in doubt. This means coal is again becoming very important as a means of meeting Britain's fuel needs.

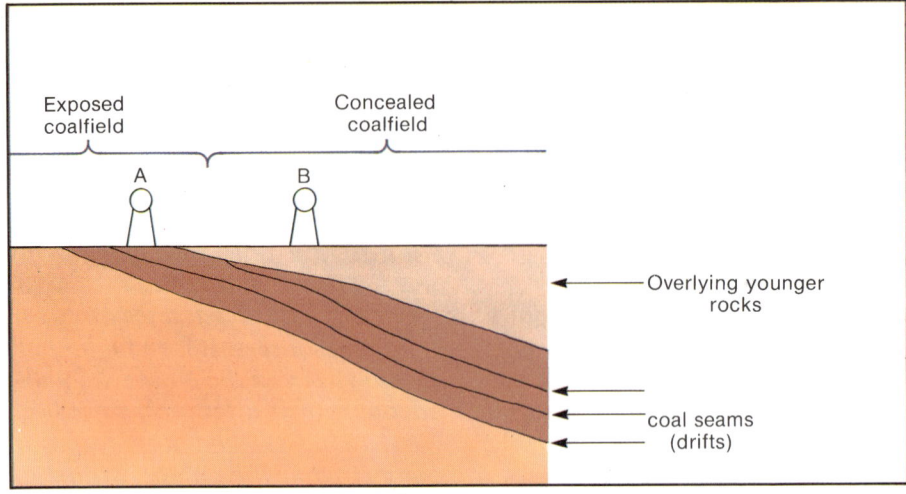

Figure 1.2. *Exposed and concealed coalfields*
Two collieries are marked A and B. Which colliery will need a shaft? Which colliery will be the cheaper to operate?

○ WHAT IS COAL?

Coal is a black rock found underground. It was formed millions of years ago. Britain looked very different then, with many large forests and swamps (see Figure 1.1). When the trees died, they fell into swamps, together with other dead vegetation. This all rotted. As time passed, other sediment formed on top and gradually compressed the decayed material into coal. This produced a seam of coal.

Several different kinds of coal are found in Britain. The main type is bituminous coal. This burns with quite a lot of smoke and is often known as household coal. Bituminous coal is plentiful. Coking coal is of better quality, and is used in the steel industry. It burns at a high temperature. Anthracite is a hard, clean, smokeless coal. Coke and anthracite are both in short supply. Great efforts are being made to find new sources.

○ HOW IT IS MINED

Some coal is found near to the surface of the ground. This is because the younger rocks above the seam have been worn away. This exposed coal is easy to mine. However, most coal is mined from concealed coalfields, deep underground (see Figure 1.2). Vertical shafts have to be sunk to reach the seams, which may be 1000 metres below the surface. The coal bearing rocks may contain between 20 and 30 seams, some up to two metres thick.

Figure 1.1. *A coal swamp*
The climate was very humid (hot and damp). Describe the scene in this photograph.

Figure 1.3. *A machine cuts coal from the coal face*

Figure 1.4. *Transporting the coal from the coal face*

Figure 1.5 *The two winding towers*
Using the information in Figures 1.3, 1.4 and 1.5, describe how the coal is mined and brought to the surface.

Deep mining produces 88 million tonnes of the coal industry's 105 million tonne output. The remaining 17 million tonnes is obtained from opencast mines (see Figure 1.6) and licensed mines.

In opencast mines, excavators remove the top soil, and then 'scrape' out the coal, which lies underneath. When the coal has been removed, the land is used for farming again, or made into a recreational area. Normally, these mines only have a short life (three to fifteen years). One of the largest opencast sites is near Morpeth in Northumberland, where almost thirteen million tonnes will have been removed between 1977 and 1990. One advantage of this method of mining is that the waste and derelict land can be reclaimed after the coal has run out.

Figure 1.6 *Getting coal by the opencast method*
The photograph shows an excavator known as 'Big Geordie'. Describe what is happening. What are the advantages of getting coal in this way? How are people affected by this method of mining?

○ WHERE IT IS MINED

The main coalfields are shown on the map (Figure 1.11). They extend from South Wales to central Scotland. The most important area lies to the east of the Pennines. Many coalfields are declining because of difficulties in mining the coal. South Wales shows this (Figure 1.7). When coal was first mined in South Wales, it was taken from the upper seams. This was very easy because the levels, or 'drifts', could be driven into the hillside horizontally. But there were dangers. What might these have been?

Most of the drift mines are now exhausted. Much of the coal now comes from the deep shafts sunk vertically at the valley floor. Today, the South Wales coalfield still suffers from mining problems and it is expensive to mine the coal. As a result, many mines have closed and the number of miners has decreased. This has caused unemployment in South Wales and has affected life in the mining communities. Many miners have moved to other areas of Britain, particularly the Midlands. They are attracted by jobs in the car industry. These jobs offer higher wages and better working conditions. However, light industries, such as washing machine manufacturing, have been brought into South Wales. This has helped stop people leaving to find work elsewhere.

○ USES OF COAL

Over the years, output of and markets for coal have changed dramatically. Figure 1.12 shows the main changes that occurred between 1949 and 1983.

Referring to the table, suggest reasons for the fall in coal production. Suggest reasons why large changes have taken place in the use of coal since 1949.

Figure 1.7. *S. Wales coal output (tonnes)*

1913	54 million
1953	24.9 million
1963	19.4 million
1973	7.3 million
1983	6.9 million

Figure 1.8. *Number of miners in S. Wales*

1953	110 100
1963	59 900
1973	30 900
1983	22 700
1985	13 500

Figure 1.9. *Production of deep-mined coal (millions of tonnes) 1985–86*

Scotland	4.3
North East	9.6
Yorkshire, Nottinghamshire and Derbyshire	51.8
South Midlands	6.3
West	9.4
South Wales	6.6
TOTAL	88.0

Figure 1.10. *Colliery output per person per shift (Figures in tonnes) 1985–86*

Scotland	9.9
North East	11.1
Yorkshire, Nottinghamshire and Derbyshire	12.6
South Midlands	11.9
West	15.4
South Wales	8.1

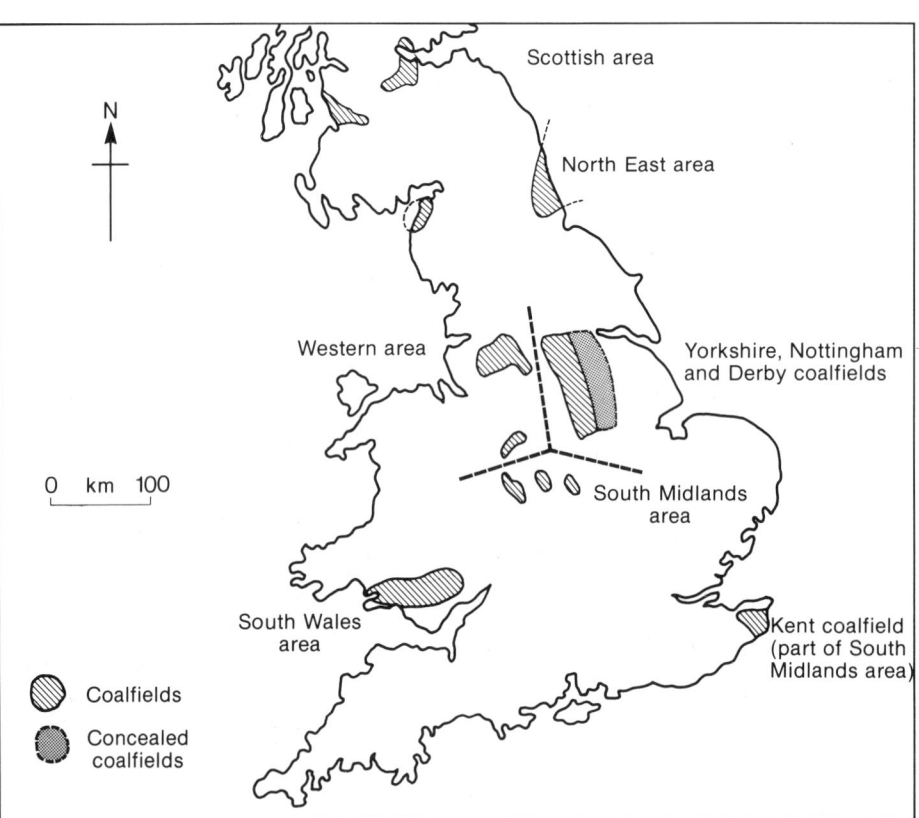

Figure 1.11. *Main coalfields in Britain*

Questions

Look at Figures 1.9 and 1.10. Then answer the following:

1 Which coal area produces:
(a) the most coal;
(b) the least coal?

2 Draw a bar graph to show the production of coal by areas. Arrange the columns with the highest on the left and the smallest on the right.

3 Which coal area has:
(a) the highest output per person per shift;
(b) the lowest output per person per shift?

Resources · COAL

Figure 1.12(a)

Year	Total coal output (million tonnes)	Number of collieries	Output per person/year (tonnes)	Percentage of output mined by machines	Number of people employed (thousands)
1953	223.5	875	301	6.0	711
1963	197.7	611	370	61.2	544
1973	138.3	281	390	93.0	268
1983	120.9	191	504	95+	191

Figure 1.12(b). *Changing markets for coal (Million tonnes)*

1949	Market for coal	1983
34	Power stations	87
27	Coke ovens	6
38	Domestic	7
42	Industry	9
15	Railways	—
27	Gas supply	—
15	Other	4
20	Export	3
218	Total	116*
	Imports	3
*This figure includes open cast mining.	TOTAL	119

Figure 1.13 and 1.14. *Typical coal mining scene in the 1920s, with the mine at the end of the street and a row of miners' houses. Describe these scenes and suggest what the area might look like today. There is high unemployment there. What can be done to overcome this?*

○ CHANGES IN THE COAL MINING INDUSTRY

Since 1945, considerable changes have taken place within the coal mining industry, as shown in Figure 1.12(a) above. As more machines are used, further changes are expected in the future.

For example, 25 collieries were closed in 1985/1986 as their total output was very low. Most mines now have an annual output of over two million tonnes.

Closing a colliery means that miners may be sent to other pits. There is often unemployment and some people retire early.

The two photographs 1.13 and 1.14 show typical scenes in the 1920s. The colliery has closed now but the houses still stand.

○ THE NORTH NOTTINGHAMSHIRE COALFIELD

The North Nottinghamshire Coalfield is in the large Yorkshire, Derbyshire and Nottinghamshire region. The coalfield is centred on the town of Mansfield and on the area to the east. The map (Figure 1.15) shows its size and position. The coalfield employs 17 000 men and produces over 11 million tonnes of coal per year. This is almost one-tenth of all Britain's coal. Unlike many other coalfields, its mines are fairly modern. Only a quarter were operating before 1900. There is much mechanisation and it has the highest output of any coalfield in Britain. Much of the coal is obtained from seams sloping very gently to the east. Some of the mines are over 70 metres deep. Most of the coal is obtained from a seam called Top Hard. This is slowly becoming exhausted and other seams, such as Deep Soft and Parkgate, are gradually being developed. This means that different types of coal can be mined for a greater variety of markets.

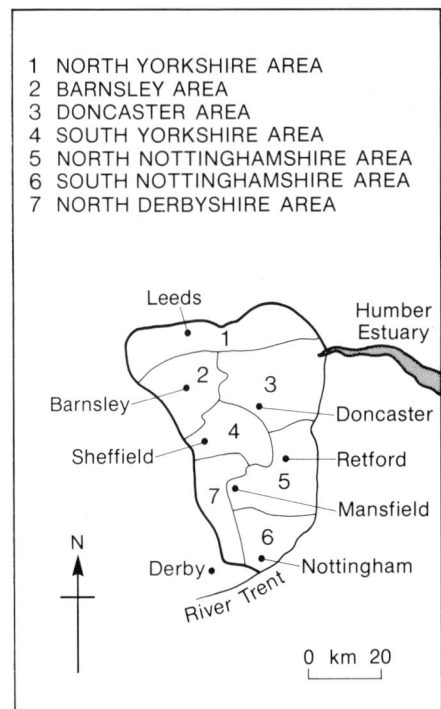

Figure 1.15. *N. Nottinghamshire Coalfield*

Figure 1.16. *Uses of Nottinghamshire coal*

Coal-fired power stations	54%
Industry	20%
Coke ovens	9%
Smokeless coal and to the domestic market	12%
Other uses	5%

○ NEW COALFIELDS

Because old mines become difficult to work, there is a constant search for new underground seams. Rich seams of coal have been discovered recently in the Selby Coalfield to the south of York and the Vale of Belvoir in north-east Leicestershire.

Rich reserves of coal were found around the Yorkshire town of Selby in 1967. Work began on sinking mine shafts in 1977. The first coal was brought to the surface in 1983.

Five mines will eventually produce coal. You can see their locations in Figure 1.19. Wistow was the first to open. The others should be in operation by the late 1980s. Together they are expected to produce 10 million tonnes of coal each year.

All the coal from the new coalfield will be brought to the surface at Gascoigne Wood Drift Mine and not at the other mines. A system of conveyor belts will carry the coal underground to this one site. Most coal will leave Selby by train, using the existing railway sidings.

This system should be very efficient. It should also help to reduce the impact of the new mining on the local environment. Waste material will be removed in the same way as the coal. This will mean no 'slag heaps'. The absence of railway sidings will also help to reduce the 'visual pollution' caused by the new collieries. Much of the coal produced will be taken by 'merry-go-round' trains to local power stations. Two are shown on the map (Figure 1.19) at Eggborough and Drax. The new coalfield will provide work for approximately 4000 people.

The Vale of Belvoir also has rich reserves of coal below the surface. Permission to mine coal here was finally given in early 1984. Many people had opposed the scheme. They believed that coal mining would ruin one of the most picturesque parts of England. They also thought many acres of valuable farmland would be threatened.

It is likely to be 1990 before full production is achieved at this new coalfield.

Bevercotes Colliery is a good example of a modern coal mine.

Much of the coal from this field goes to the power stations along the River Trent. Other markets are shown in Figure 1.16. Several mines operate 'merry-go-round' trains to the power stations. One train also supplies coal from Welbeck Colliery to a large cement works in south-east England.

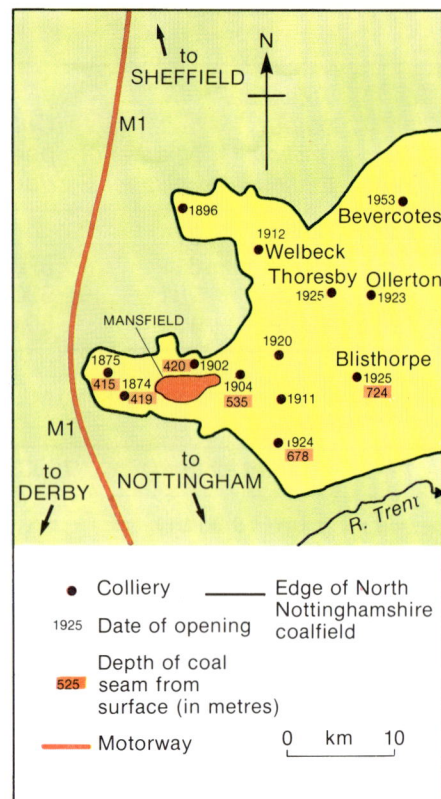

Figure 1.18. *North Nottinghamshire coalfield*

Figure 1.17. *Bevercotes colliery, a typical coal mine in Nottinghamshire. Describe how the buildings in the photograph might be used.*

Questions

1 Look at Figure 1.18, which shows the depth of the main coal seam and the ages of the mines in different collieries. What do you notice about the depth of the seam and the age of the colliery? What does this tell you about the seams of coal underground?

2 A new mine is to be built in this coalfield. You are responsible for deciding where it should be located. Choose a suitable location, and give reasons for your choice.

3 When your mine is eventually built, the scene will be like the one in the photograph (Figure 1.17). Local people are likely to object to having the area spoilt in this way. You will have to justify the building of the mine, so
(a) prepare reasons for wanting to put a coal mine there;
(b) think of the arguments that might be used by people who do not want a coal mine built;
(c) think what sort of groups might oppose a mine.

Resources · COAL

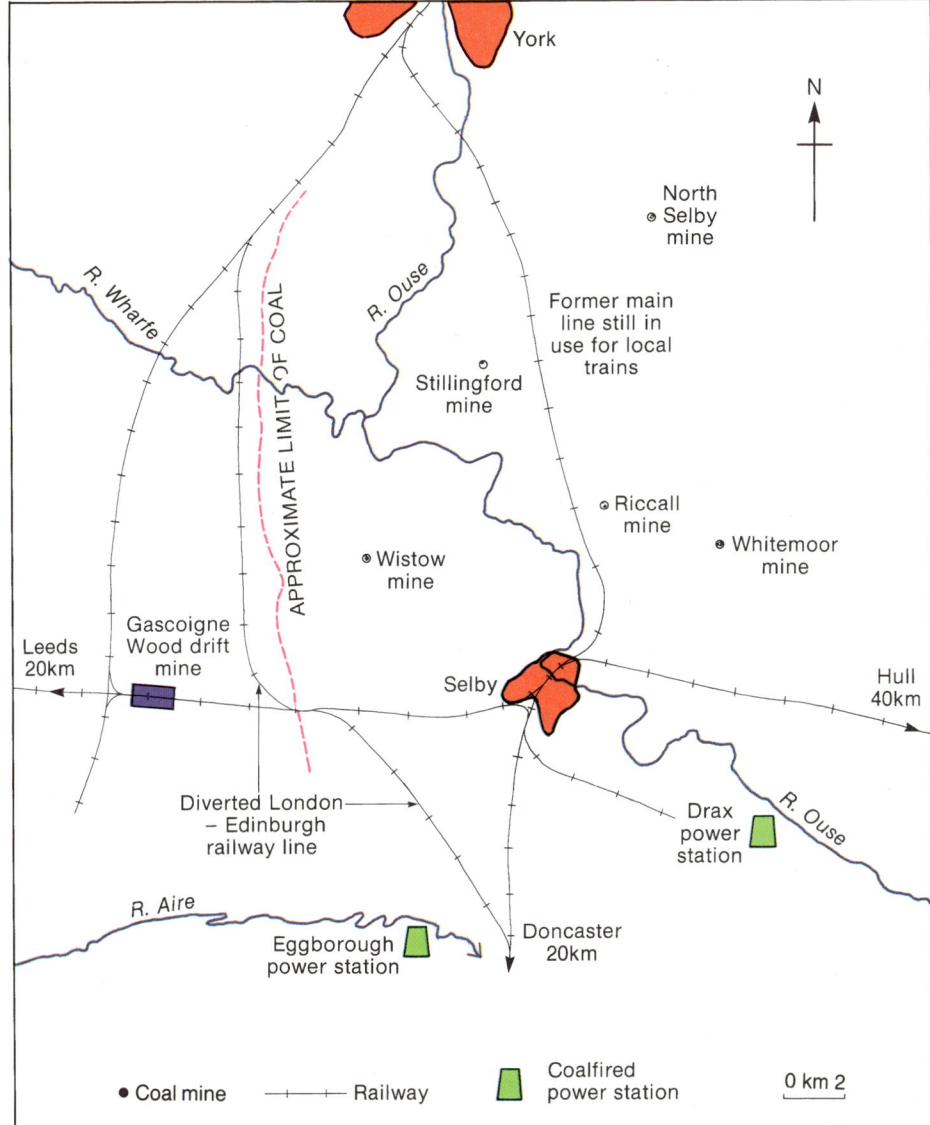

Figure 1.19. *The Selby coalfield, showing the five coal mines*
The former railway line through Selby will suffer from subsidence as the coal is mined. As a result, a diversion has been built to allow High Speed Trains to operate.

Figure 1.20. *Wistow mine buildings.*
Look at Figure 1.19 and find where Wistow mine is located.

Questions

1 Which people are going to be most affected by the development of the Selby coalfield?

2 Why was it decided to bring all the coal to the surface at Gascoigne Wood Drift Mine?

○ PROBLEMS IN COAL MINING

One of the problems in mining is deciding what to do with the waste. In the past, mines dumped their waste material in conical mound 'slag heaps' near the colliery. These mounds are unsightly features of the landscape in old coal mining areas. They can also be dangerous if they slip down hillsides. This happened in Aberfan, South Wales, in 1966. The huge slag heap behind the village suddenly became waterlogged and slid down into the valley. Houses and the local school were buried. Five teachers and 116 children were killed. In recent years many slag heaps have been flattened and grassed over. Some of the reclaimed land has been used for housing development.

Questions

1 Where could the waste material from mines be taken, instead of making conical mounds?

2 What other problems are there in coal mining areas? How might these be overcome?

3 What evidence is there to suggest that South Wales has the largest problems of all the coalfields?

Gas and Oil

In 1965, natural gas was found under the North Sea, off the coast of East Anglia. This was the beginning of a new and exciting venture for Britain. Companies started to drill for more gas. The large deposits they found were called 'gas fields'. Later, the search spread northwards and large quantities of oil were found.

Both gas and oil are found in rocks over 250 million years old. Both formed from decaying animals and plants which were living when the rocks were laid down. During the millions of years that followed, the rocks became folded. The gas and oil migrated to a rock that they could pass through easily. Sandstone was an ideal rock for this purpose. But as long as a cap of a rock such as clay was overlying the sandstone, the oil and gas could not escape.

Before the gas could be brought ashore by pipeline, there were three important stages:
(a) *Seismic searches* These involved a ship sending down small shock waves to the rocks beneath the water. This told geologists whether there was any likelihood of gas being present. If the area looked promising, licences to drill were bought.

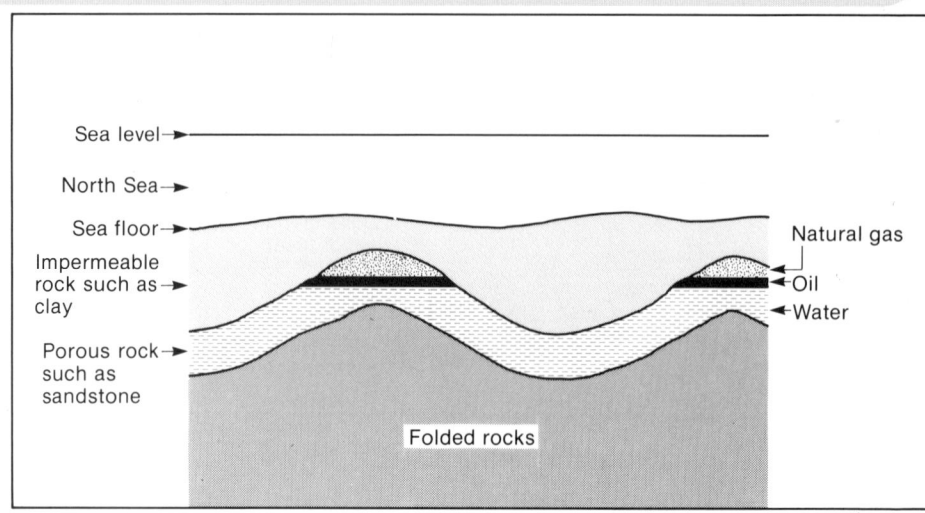

Figure 1.22. *A section through part of the North Sea, showing oil traps*
Describe why the oil and gas are found here.

(b) When a licence had been bought, mobile drilling rigs were towed in to test whether there actually was any gas. Several wells were sunk to discover the amount of gas present. It might cost £20 000 a day to hire a rig, and over £10 000 000 to find out whether there was enough gas to be worth going into the third stage.
(c) If large quantities were found then a permanent platform was constructed. The gas was then brought to the surface and pumped ashore (see Figure 1.21).

In 1967 the first natural gas was piped ashore from the West Sole gas field. The following year, gas was piped from the Leman gas field. Ten gas platforms have been built on this field since. The gas arrives at three gas terminals on the east coast of England. Gas is then distributed around the country by underground pipes laid out in the form of a grid.

After gas had been found in the southern part of the North Sea, the search moved northwards into the area between Scotland, the Orkney Isles and Norway. Here large quantities of oil were found, together with small amounts of gas.

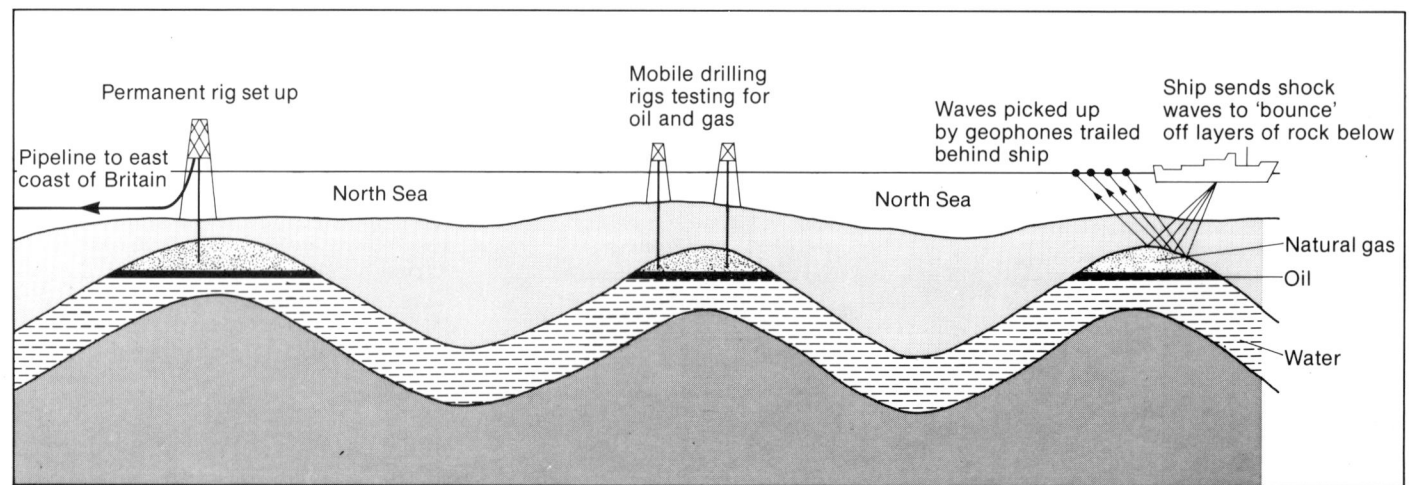

Figure 1.21. *The search for oil*
Describe the way in which oil is located under the North Sea.

Resources · GAS AND OIL

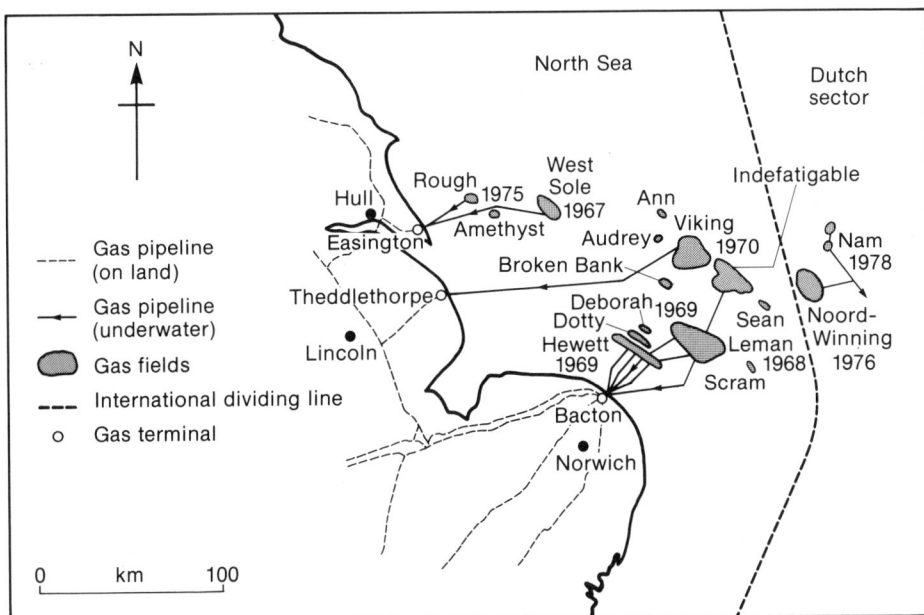

Figure 1.23. *The southern part of the North Sea, showing the main gas fields*
The gas fields in operation have the date when production started. When did most fields in this area start production? Using the scale, work out the sizes of the main fields.

In 1971, the first oil was brought ashore from Ekofist in the Norwegian sector. Four years later, the large Fortes field started to provide Britain with oil. After that a large number of oil fields were developed in a similar way to the gas fields. Figure 1.24 shows the present extent of proved oil fields in the North Sea. Oil has been discovered also in numerous other areas.

When sufficient oil has been found, a 'platform' is built to organise and control the amount of oil taken out of the rocks. These platforms are very large and provide homes for the workers extracting the oil. The workers live on the platform for up to several weeks at a time. They need to be provided with a restaurant, sleeping quarters and recreation areas. The platform is like a small hotel.

It is important that the platforms do not move. They have to withstand the strongest gales in the North Sea. Some platforms are anchored to the sea floor by thick cables. Others are actually fixed to the sea floor by four giant legs. Any slight movement could cause an oil leak and a lot of pollution. Some of the oil is transported ashore by tankers. The rest is pumped to terminals like Sullom Voe.

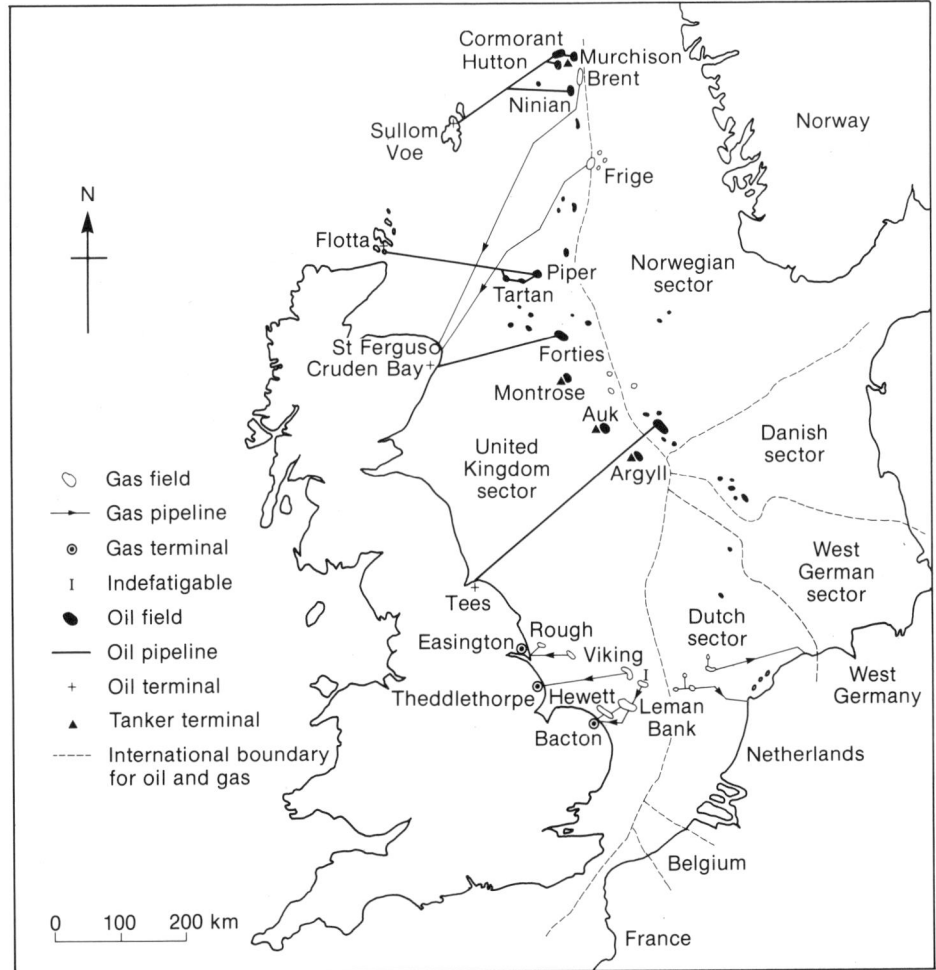

Figure 1.24. *Oil and gas in the North Sea*
The main oil and gas fields are marked and named, but there are many others that have been discovered and are not in production yet. Some oil fields have tanker terminals alongside. Describe what these might be like. Can you suggest why tanker terminals are chosen rather than building a pipeline?

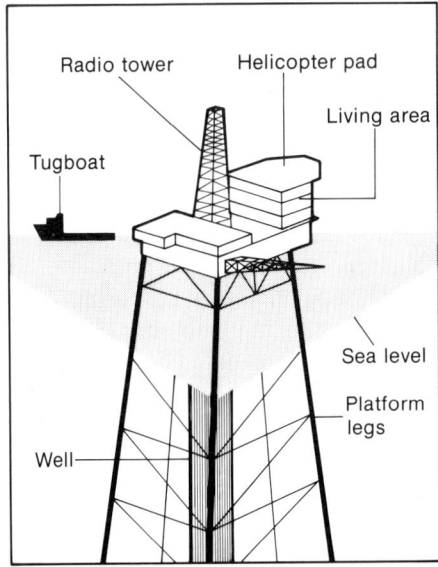

Figure 1.25. *An oil platform*

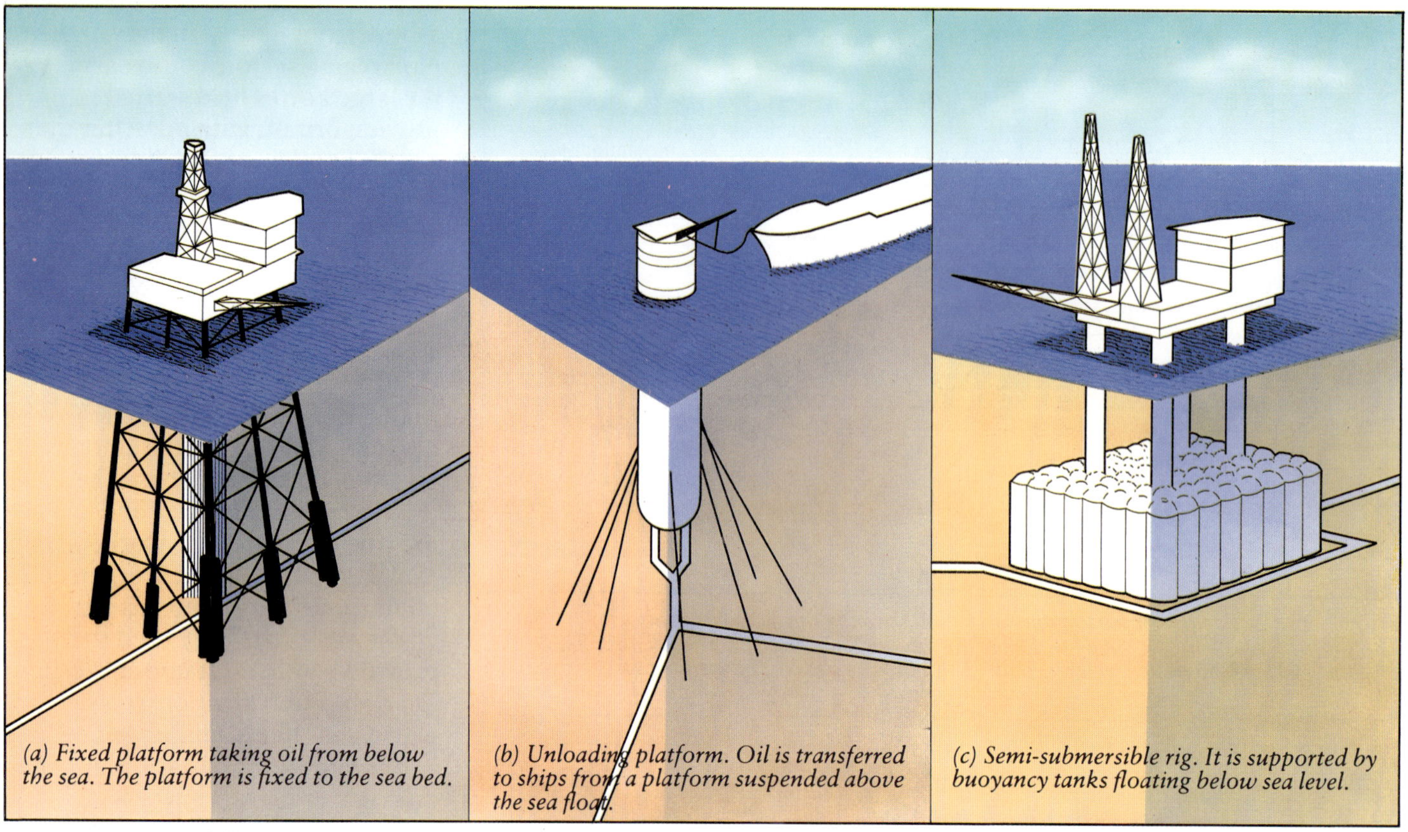

The British Isles: Themes and Case Studies

(a) Fixed platform taking oil from below the sea. The platform is fixed to the sea bed.

(b) Unloading platform. Oil is transferred to ships from a platform suspended above the sea floor.

(c) Semi-submersible rig. It is supported by buoyancy tanks floating below sea level.

Figure 1.26. *Three different types of anchoring platforms*

○ SULLOM VOE

Some of the oil from the northern oil fields of the North Sea is brought ashore by pipeline to Sullom Voe. Figure 1.24 on page 9 shows that Sullom Voe is situated in the remote Shetland Isles. The oil is stored in large tanks until it can be collected by ships. It is then taken to refineries around the coast of Britain.

Sullom Voe was chosen as the site for a large oil terminal for a number of reasons:

(a) It is the nearest suitable area of land to the oil field.
(b) It has a sheltered deep water channel. This means that oil tankers can berth very close to the oil terminal. These tankers can be very large, capable of carrying up to 300 000 tonnes of oil. The largest tankers are known as ultra large crude carriers (or ULCCs).
(c) The land is fairly flat. This makes it easier to build large oil storage tanks.

Figure 1.27. *Sullom Voe*

Resources · GAS AND OIL

Before the oil terminal was built, crofting was the main local occupation. There were few other jobs. Many people had to leave the Shetlands to find work elsewhere. Building and running the oil terminal has provided much needed local employment for the islanders.

Sullom Voe is very remote and barren. All the materials needed to build the terminal had to be brought across the sea from Scotland. This was difficult and expensive. The land was also covered with peat. Large areas had to be drained before any building could begin.

Sullom Voe handles over one million barrels of oil a day. (There are 7.4 barrels in a tonne of oil.) The oil terminal can handle more than this if the need arises. A large number of oil companies use Sullom Voe, but it is actually owned and operated by the local authority.

With so much oil going in and out of the oil terminal, many people worry about the possibility of oil pollution in Sullom Voe. They fear that oil tankers will spill oil or that storage tanks will leak. Oil spillage badly affects wild life, often killing large numbers of birds and fish. If a bad spillage occurred, the coast around Sullom Voe would be heavily polluted. This would have a serious effect on the remaining crofters. One incident has already caused problems. In 1979 a tanker collided with an oil jetty and over 1000 tonnes of oil poured out.

Many Shetlanders are also worried about the future of the Islands for a different reason. Now that the terminal is complete, many of the 6000 workers have left Sullom Voe. This has caused problems in the community. These workers spent money on local goods and services which kept many shops and small companies in business. Employment and social problems are becoming worse than they were before the terminal was built.

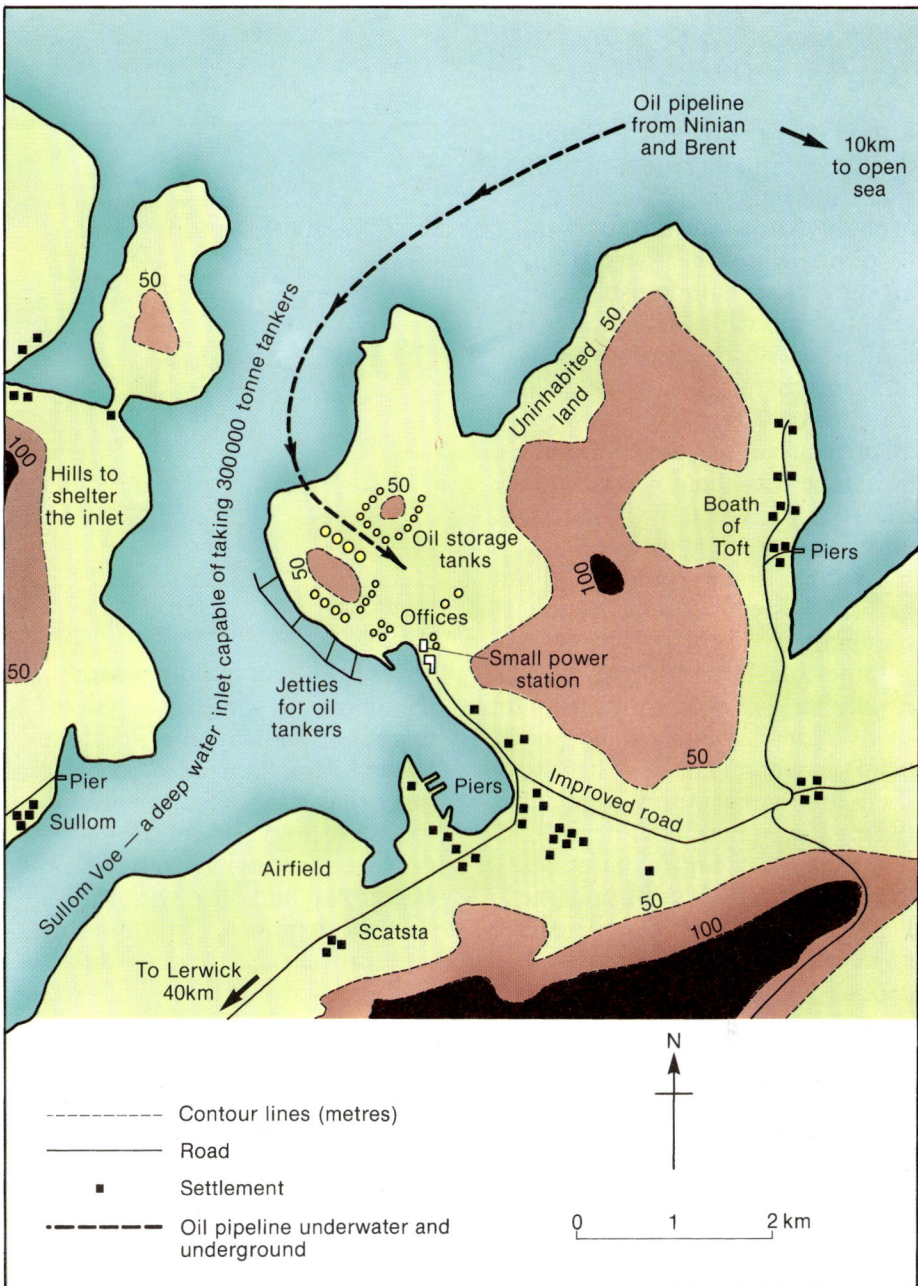

Figure 1.28. *Sullom Voe in the Shetland Isles is one of the most recent oil terminals Using the information on the map, state the advantages of building an oil terminal there. Describe the location of the settlements. What occupations might the people living there have?*

Questions

1 Briefly describe, with the aid of diagrams, how oil and gas forms in the rocks.

2 The search for oil and gas costs a lot of money. Explain why.

3 What are the main problems of obtaining oil and gas from the North Sea?

4 Some of the oil is transported ashore by tankers and some by pipeline. What are the advantages and disadvantages of each method?

5 What are the main problems of having an oil terminal at Sullom Voe? How might they be overcome?

Limestone

Britain is in an enviable position in that it contains a great variety of rocks and minerals. You will probably know one type of limestone which weathers to produce very characteristic features. It is called Carboniferous Limestone. 'Carboniferous' indicates that the limestone contains carbon. Coal is largely made from carbon too, but the limestone was laid down before the coal formed. When the limestone is pure, it weathers to form spectacular underground caves and caverns – as well as stalactites and stalagmites.

Limestone is made up of calcium carbonate. The rock forms from the shells of dead animals and deposits from sea water. The deposits and shells are then compressed under the great weight of later deposits.

Limestone occurs in many locations throughout Britain (see Figure 1.29). It is only quarried in areas where there are extensive deposits. Some of the quarries are huge. They may make ugly scars on the landscape. Large quantities of fine white dust are produced which cover fields and buildings around the quarry. In some areas where there is no railway, the limestone has to be transported away by lorry. This can be dangerous if the roads are not very wide. Some disused quarries have now become overgrown and filled with rubbish.

In the Pennines, Carboniferous limestone is used for dry stone walling. Many of these walls were originally built out of local stone, hundreds of years ago. The remains of many small disused quarries can be seen dotted all over the local landscape.

Limestone is used as a raw material in the chemical industry and in cement making. Massive quantities are needed for these industries.

In the Cotswold Hills, another type of limestone occurs. It is called Jurassic Limestone or 'oolitic limestone'. It is not as hard or as pure as some of the Carboniferous limestone, but it is used for building local houses.

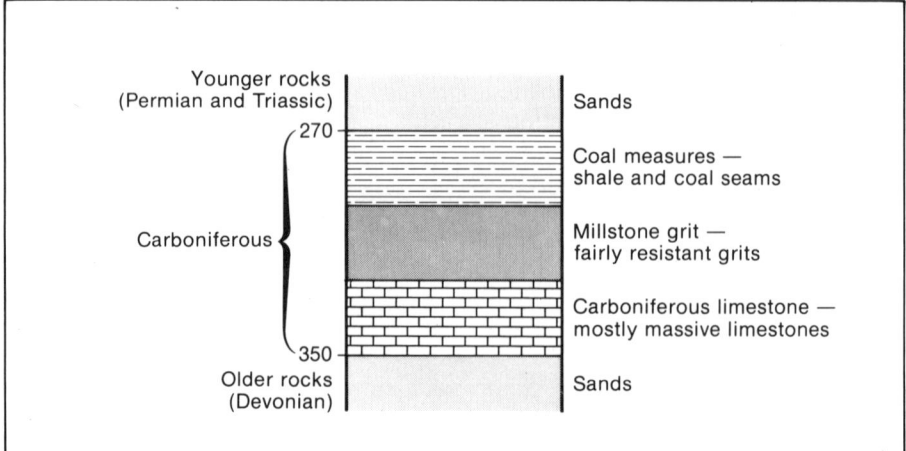

Figure 1.30. *Part of the Geological Column, showing the order of beds. The figures represent the age of the rock in millions of years.*

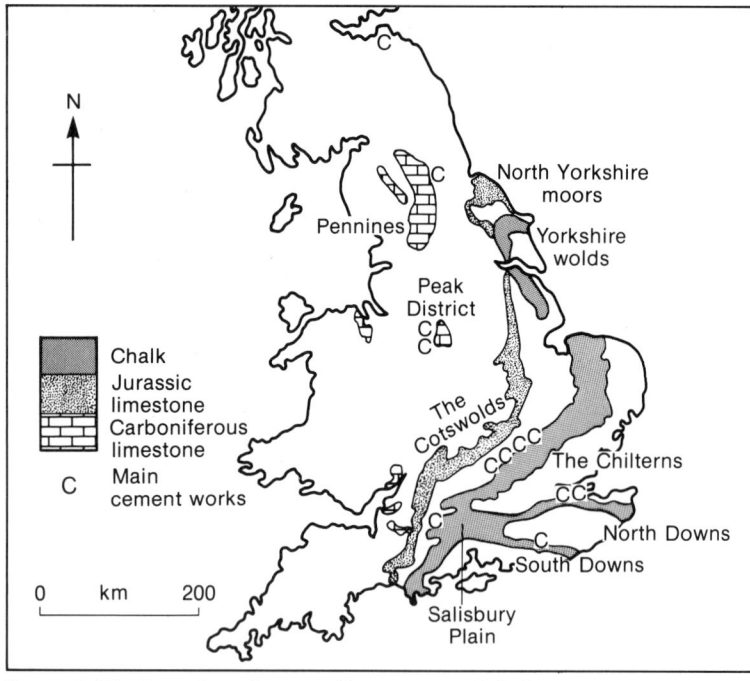

Figure 1.29. *Location of areas of limestone and chalk*

Figure 1.31. *A building in Bath made from oolitic limestone. The stone is used for many buildings in the city.*

Resources · LIMESTONE

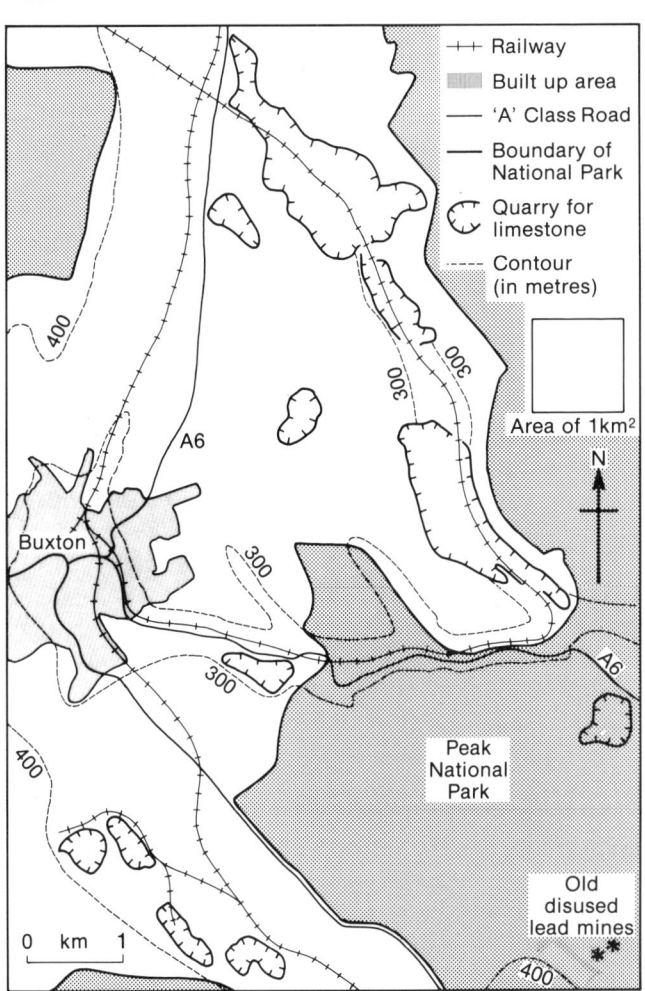

Figure 1.32. *Distribution of limestone quarries around Buxton in the Peak District*

Figure 1.33. *Magnesium limestone. Various types of limestone are the most-quarried rocks in Britain today.*

Figure 1.32 shows the area around Buxton, in Derbyshire. Buxton is a market town in the Peak District, on the edge of the National Park. It is a popular resort with hotels and spa waters. To the east and south there are extensive limestone quarries, many of which are being worked at present.

○ CHALK

A special type of limestone is called chalk. When pure, it is soft and white. It is made largely from the remains of minute sea creatures. Most chalk is very pure. It is made up of calcium carbonate.

As chalk is soft, it is not used as a building stone. However, chalk contains flints. These are hard nodules of silica. In the south of England, flints are used to build walls and to decorate the sides of houses. Flint is also used as a hard core for the foundations of roads and houses. Some flint is used in the pottery industry at Stoke-on-Trent.

Chalk is an important raw material in the manufacture of cement. Like other limestone, chalk is quarried from hills which are easily accessible. These large quarries are very unsightly. In the south of England, many parts of the chalklands are considered 'areas of outstanding natural beauty'. There is conflict between companies operating the quarries and conservationists.

Even with all the quarrying, both limestone and chalk areas have much to offer the tourist industry. Quarrying needs to be developed very carefully.

Questions

Study Figure 1.32 which shows the site of Buxton and the neighbouring limestone quarries, and then answer the following questions.

1 What is the length and area of the largest quarry marked on the map?

2 Describe and explain the routes taken by the railway lines shown on the map.

3 Explain why the railway is important for moving limestone.

4 What do you notice about the extent of the quarries compared with the boundary of the National Park? Why do you think this has occurred?

○ CEMENT MAKING

Cement is used in the construction industry for buildings, bridges, and, in some cases, roads.

The raw materials used in cement making are limestone or chalk, clay and gypsum. These raw materials are bulky. As transport costs are high, cement making tends to be located where good supplies of chalk or limestone can be found locally.

The map of England in Figure 1.29 on page 12 shows that the cement industry is located on, or near, limestone or chalk rocks. The main areas are in:
(a) Derbyshire, where Carboniferous limestone is found;
(b) the Cotswold Hills, where Jurassic limestone occurs;
(c) south-east England, where chalk rock is found in the North and South Downs.

The Midlands and London provide a ready market for the cement. Figure 1.34 shows one cement works in south-east England. This is the Shoreham Works in Sussex. There is a ready supply of chalk from the South Downs around the works. The photograph shows how close the quarries are.

Clay is brought from a large clay pit north of the works. At the pit, the clay is mixed with water to form a slurry. This is then pumped to the cement works by underground pipeline. Any hard nodules or large fossils are taken out of the slurry before it leaves the clay pit.

The other raw material needed is gypsum. Smaller quantities of gypsum than chalk are used, so it is not too difficult or expensive to bring it from further away. Most of the gypsum used at Shoreham comes from Robertsbridge in East Sussex, a distance of 70 km.

Large amounts of water are needed to make cement. The River Adur provides all that is required. In the past, the river was also needed to transport the finished cement to the harbour at Shoreham. In return, coal was brought back from the port to be used in the cement works' boilers. Today the works are powered by electricity.

Trains were also used to transport the finished cement. The railway line which passes close to the works has now been closed. This means that most of the cement produced at Shoreham is carried to customers by road.

When cement making began at Shoreham, electric drills were used to dig the chalk out of the Downs. Now the quarry has become much larger so the chalk has to be blasted free. It is then taken the short distance to the works by lorries and dumper trucks. When it arrives, the chalk is crushed and filtered to remove the stones and flint. The flint is used in the pottery industry at Stoke.

Today, the industry at Shoreham makes up to 370 000 tonnes of cement each year. Much of the finished product is used in the south-east and south of England.

Figure 1.34. *The Shoreham Cement Works in Sussex. Describe the scene in this photograph.*

Resources · LIMESTONE

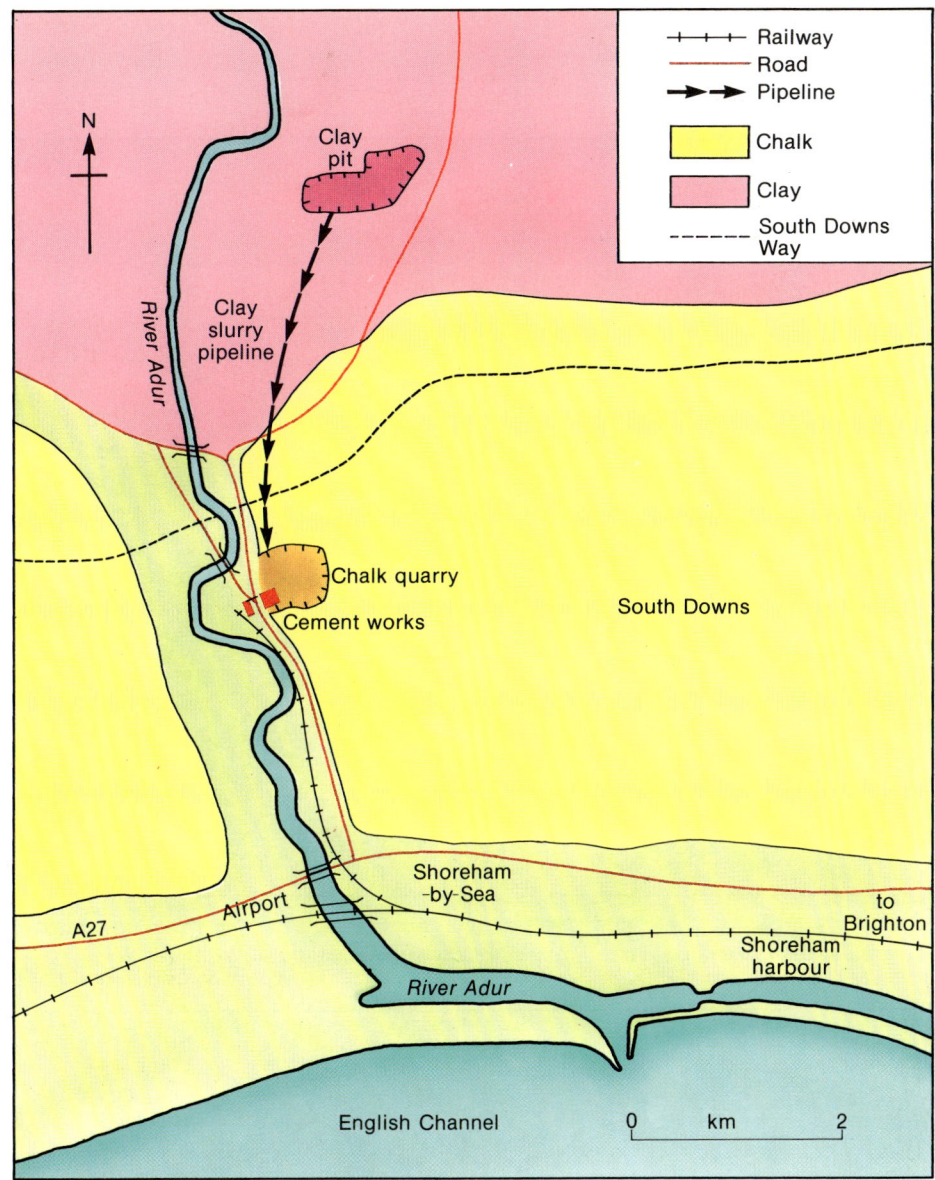

Figure 1.35. *Location of Shoreham Cement Works*
The map shows the position of the main raw materials and transport routes.

A large chalk quarry is an ugly blot on the landscape, particularly in the beautiful South Downs. Planning has to be carefully controlled, as there are fears that the quarry may become considerably bigger. Until the 1950s, dust was another problem for people living near the cement works. It used to blow over the surrounding countryside and cover everything with a white film. The problem has been reduced in recent years. New, expensive equipment has been installed which reduces the pollution. Today, surrounding fields are green again and the roofs of the houses shown in the photograph (Figure 1.34) are no longer covered with dust.

A third problem existed until 1980. The road through the works was very dangerous because it was narrow and twisting. All the lorries entering and leaving had to join the road at its narrowest point, on a bend. A new road has now been constructed through the side of a hill. The chalk taken out during the building was used to make cement. Both the old and the new roads can be seen on the photograph (Figure 1.34).

The cement works employs 330 people. Some live in houses near the site. Most workers, however, live in the towns and villages surrounding Shoreham.

Questions

1 With reference to Figure 1.35, describe and explain the advantages this area has for the production of cement. Include in your answer: raw materials, transport, markets, labour supply.

2 Explain the main objections that would arise if the cement works expanded.

Figure 1.36. *The chalk quarry at Shoreham Cement Works*

Water

Every time we wash, drink, use the garden hose or wash the car, we take it for granted that water will come pouring out of the tap. Few homes have water meters so most of us have little idea of how much water we use. In fact, each person in Britain uses an average of 120 litres (about 27 gallons) each day. This includes 35 litres a day on washing and taking baths. Another 11 litres goes down the drain after washing up and general cleaning. If you decide to hose the garden, that can use up to 900 litres of water an hour.

Although a great deal of water is used in the home, industry demands much larger quantities. For example, 115 litres of water are needed to make 1 kg of powdered milk and 26 000 litres are used in the production of one tonne of newsprint. It takes 45 000 litres to make one tonne of steel.

Before 1974, most of Britain's water was supplied by a large number of small water companies. Each company had its own water supply and was responsible for local sewage disposal. There was little cooperation between the different water companies and no national planning to meet Britain's water needs. This was changed in 1974. All the water companies were joined together to make ten regional water authorities. These authorities have to make certain that Britain has enough water resources for future needs. They also have to ensure that all water is treated and supplied regularly to homes and industry.

In addition, the authorities have these other responsibilities:

(a) sewage disposal,
(b) land drainage,
(c) flood control,
(d) pollution control,
(e) fisheries,
(f) recreation and amenities.

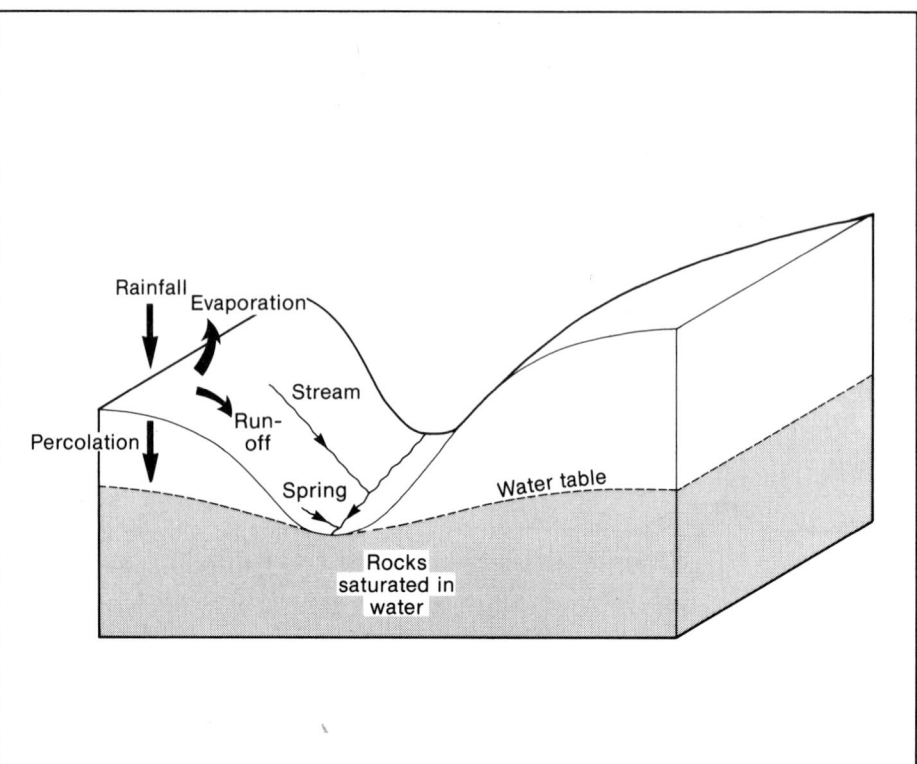

Figure 1.37. *A simple water cycle*
Rain falling on hills will go in one of three directions. What are these directions? Write a sentence to explain each.

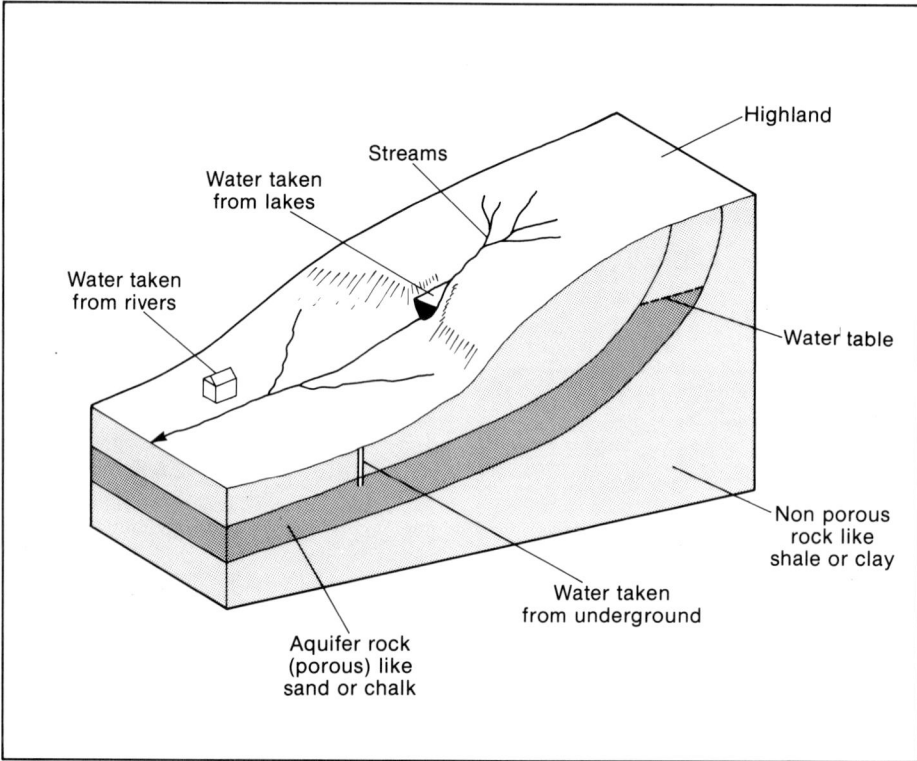

Figure 1.38. *Sources of water*
Britain obtains its water supply in several different ways as shown above. Briefly describe these ways.

Resources · WATER

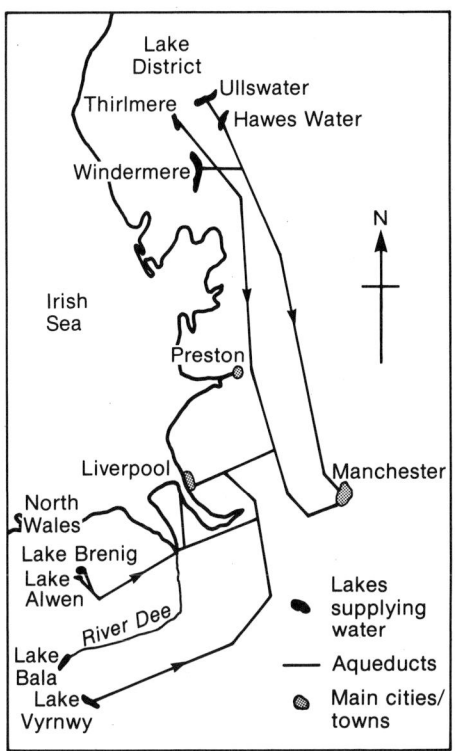

Figure 1.39. *Water supply in north-west England*

Water is obtained from four sources: underground (boreholes and wells); springs; rivers; and reservoirs. Underground water comes from rock layers which allow water to accumulate. These are known as aquifers. Water is obtained from aquifers by sinking boreholes or wells. Such water is usually very pure as it has been filtered during its passage through the rocks.

Underground water obtained from springs is also usually very pure. However, springs can only supply small amounts of water.

Some water is provided by rivers, but many rivers are polluted. Some are too dirty for their water to be used for drinking. However, as Britain's water needs increase, more water is being obtained from the less polluted rivers. It has to be purified before it is fit to drink.

Many reservoirs have been built to store rainwater. They are found in valleys where they can collect the rainfall flowing off the surrounding hills. Many are located in hilly areas where the annual rainfall exceeds 1000 mm. Rainwater, like underground water, is usually very clean, and needs little purification.

The following examples show how water is obtained in different parts of Britain.

North-west England

This region includes the industrial areas of Liverpool and Manchester (see Figure 1.39). Industry and domestic customers use about 2535 million litres of water every day. About 25% of this comes from reservoirs in North Wales, such as Lake Vyrnwy. Another 22% is supplied by reservoirs in the Lake District, such as Windermere, Thirlmere, Ullswater and Hawes Water.

More water will be needed in the future. Three possible schemes are being studied:

(a) Raising the level of some lakes in the Lake District;
(b) Sealing off Morecombe Bay and making a freshwater lake;
(c) Establishing more reservoirs, but this would mean drowning agricultural land.

The Thames Basin

The Thames Basin is controlled by the Thames Water Authority. Almost half of London's drinking water comes from the River Thames. The River Lee (a tributary of the Thames) provides another 10%, while the remainder is obtained from underground sources. The main reservoirs are situated to the west and north east of London. Staines is one example. The Thames Authority has to supply 3500 million litres of water every day. This region uses more water than any other area of Britain.

Looking back at the example of north-west England will remind you that 25% of that region's water is obtained from North Wales. The Midlands also imports large quantities from central Wales. The Welsh would like payment for all the water they export. Do you think payment should be made?

A total of 35 000 million litres of water is used every day in Britain. Figure 1.40 shows who uses it.

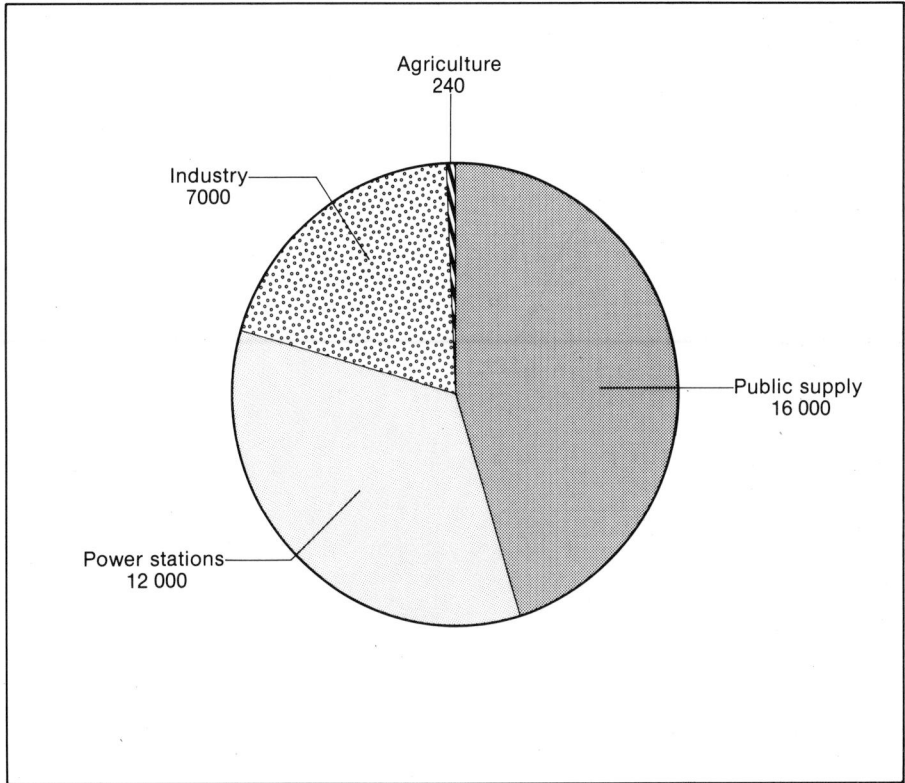

Figure 1.40. *The main uses of water*
All the figures are in millions of litres used every day. The water supplied to industry does not need purification. However, industry does take much purified water from the public supply.

○ WATER FOR THE FUTURE

In 1976, there was a long, hot and very dry summer. Much of Britain ran short of water and a drought was declared. More recently, in 1984, the south-west of England ran very short of water.

Everybody was encouraged to use as little water as possible. Hosepipes were banned. Gardens wilted, lawns turned yellow, and many trees died. Less water was used but there were still shortages in many areas.

The south-west suffered particularly badly. Much of the region's water comes from rivers and streams. These are normally low in the summer anyway. In 1976, many virtually disappeared. Large numbers of tourists to the area added to the problem by increasing the demand for water.

Since 1976, great efforts have been made to build new reservoirs. The Colliford Scheme in Cornwall (Figure 1.41) is one of five projects completed in the 1980s.

Colliford, like other schemes, needs a lot of land. A dam 8 metres high holds back a lake 5 km long and 2 km wide. The dam is 700 metres wide. It is filled with large amounts of waste sand from nearby china clay workings.

There is always opposition when land is used in this way. Some people argue that the natural beauty of the countryside is destroyed. Others claim that valuable farmland is lost, although very little of the land used in such schemes would be useful to farmers. New reservoirs have several advantages apart from providing water. They can be used for recreation (picnic areas, walks and fishing). Several lakes are used as trout fisheries.

A 'National Grid for Water' has been proposed to stop areas running short of water. This means areas of high rainfall (over 1000 mm) could supply areas of low rainfall (below 650 mm).

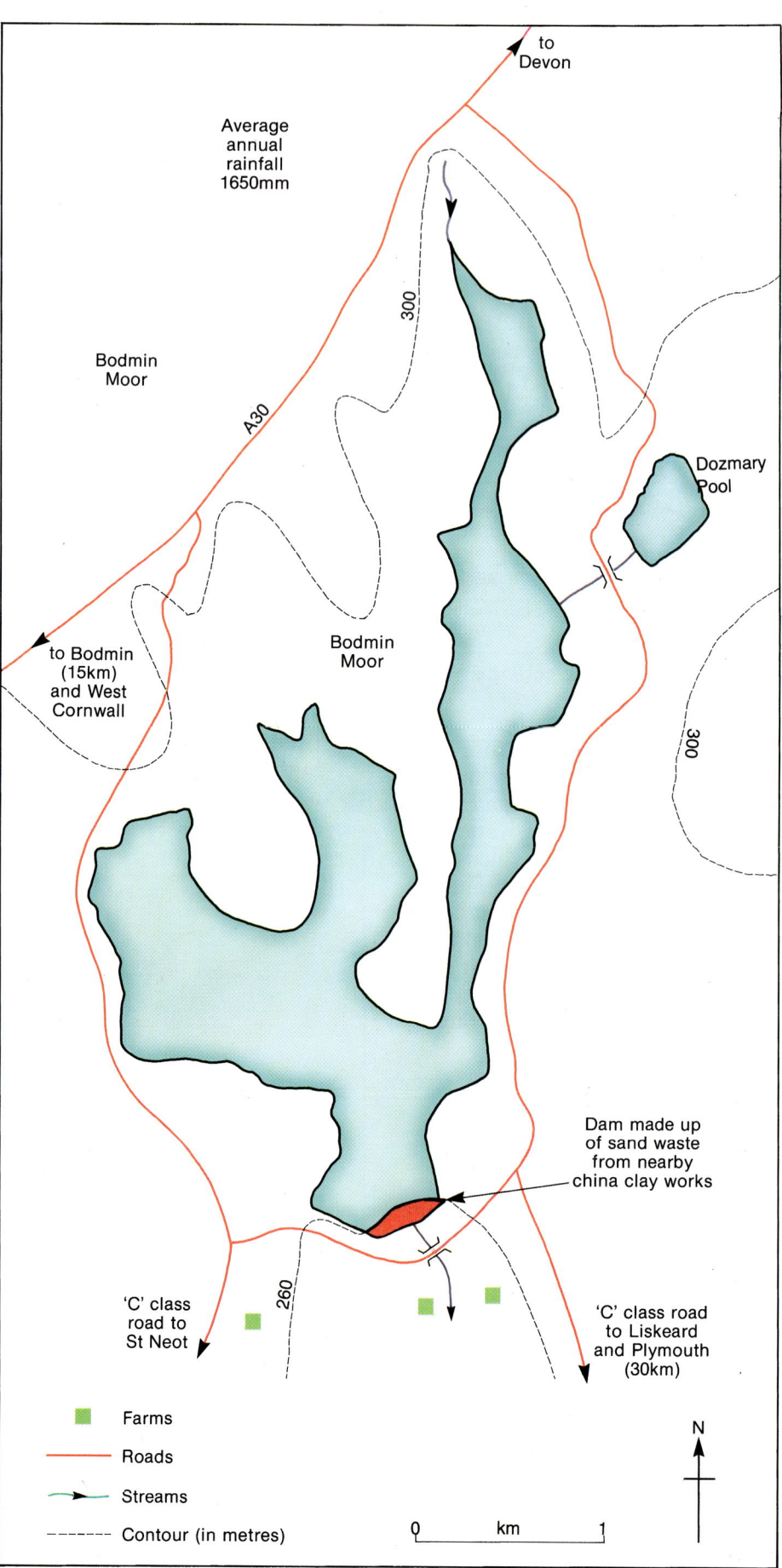

Figure 1.41. *Colliford Dam in Cornwall*

Resources · WATER

Figure 1.42. *The dam at Colliford*
Is it worth losing some farmland to have such a reservoir?

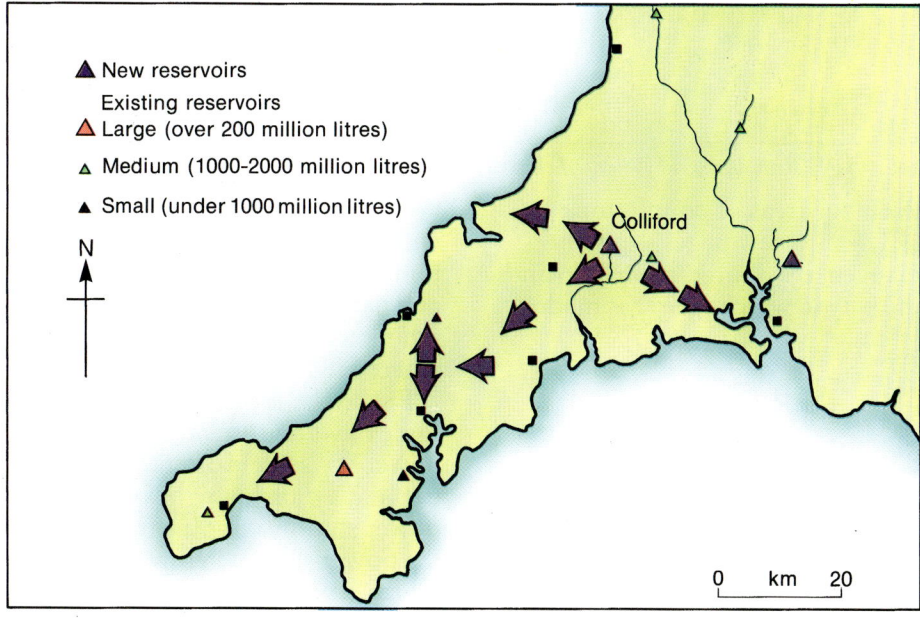

Figure 1.43. *Water for Cornwall (arrows show which way the water flows).*

Questions

1 Describe and explain the information shown on the pie diagram in Figure 1.40.

2 Describe and explain what you think the Water Boards need to do to maintain the following:
(a) sewage disposal;
(b) land drainage;
(c) flood control;
(d) pollution control;
(e) fisheries;
(f) recreation.

3 Explain why there is a need for more reservoirs.

4 Study both the map and photograph of Colliford. Suggest reasons why this area was chosen for the new dam and reservoir scheme.

5 Study the map of Cornwall in Figure 1.43. Describe where the water from Colliford will go eventually.

6 Using the scale on the map of Colliford (Figure 1.41), state the length and width of the lake and the length of the dam.

7 Using the example of Colliford, state who might have objected to the building of the reservoir. Give possible reasons for their objections.

8 Write an essay on the ways in which water is important to people.

9 Central and North Wales have a high rainfall. The water could be used to supply the London area which has a low rainfall. Imagine you were in charge of this project. These are some of the questions you would need to answer:
(a) Would new reservoirs be needed in Wales?
(b) How would the water be moved?
(c) What benefits would this bring?
(d) What problems would it cause?
(e) Would Wales want to be paid for exporting its water?
Discuss the benefits and difficulties of such a project.

Energy

Every time we switch on a television, turn on a gas fire, or make a journey by car, we are helping to use up Britain's energy resources. Large amounts of power are used in our homes. Industry and transport consume even greater quantities of gas, coal and oil.

About 250 years ago, Britain's smaller population relied mainly on wood and water for fuel and power. As the population grew, and supplies of wood began to run out, people began to look for a new fuel. They discovered that coal could provide power for machines, could be burnt in the home, and could be used as a fuel for steam powered trains. It was also discovered that new sources of power could be produced from coal. Electricity and gas produced from coal became very popular. A growing population needed more power. People began to expect, and use, more energy.

Today, newer sources of energy are also used. For example, oil is used to produce petrol, and uranium is used to produce nuclear power. We have come to expect plenty of power and fuel for our houses, industry and transport. As the population increases, we shall demand more energy. Our present energy resources are unable to meet this need. We shall have to increase our sources of power.

Coal, natural gas and oil supplies are found only in certain parts of Britain. Power stations producing electricity can only be sited in certain areas. Various ways have to be used to transport the power and fuel from where it is available to where it is needed. Grid systems are used to supply electricity and gas. Road, rail and sea transport carry supplies of coal and oil throughout Britain. Figure 1.44 shows the electricity supergrid in Britain. What do you notice about the network of the grid compared with the main areas of population?

The next few pages show the main energy sources in Britain and the way in which power is made.

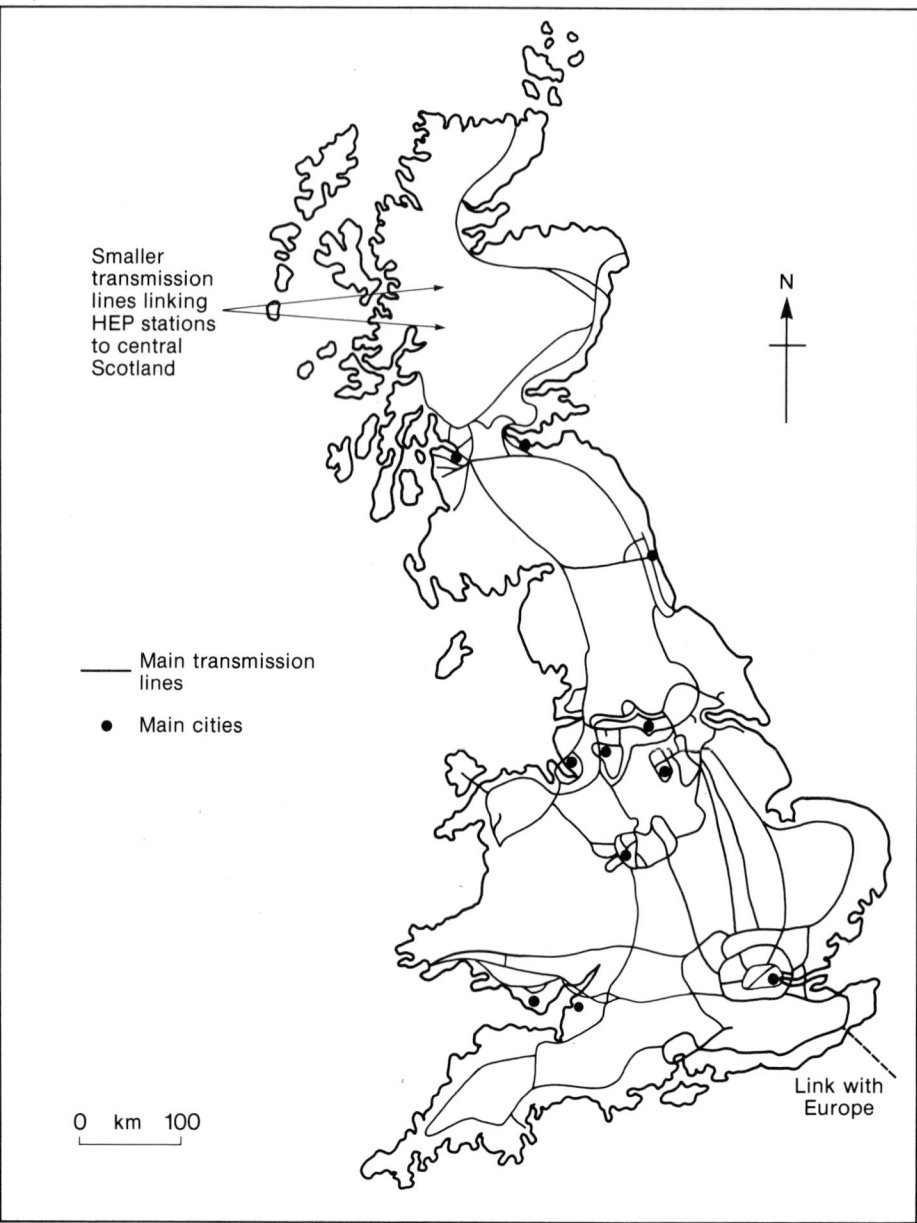

Figure 1.44 *The supergrid system of electricity*

○ POWER FROM WATER

Electricity can be produced by the force of water flowing downhill. This is called hydroelectric power or HEP. HEP has several advantages over other methods of producing power. It is clean, does not involve dangerous fuels, and will not run out. HEP is non-exhaustible, unlike coal, oil and natural gas. However, there are not many suitable places in Britain where HEP can be produced.

Electricity is produced when the force of water turns turbines in a power house. It is often convenient to build a dam high above the power station and pipe the water to the power station below. Most HEP schemes now involve building a dam with a reservoir behind.

Resources · ENERGY

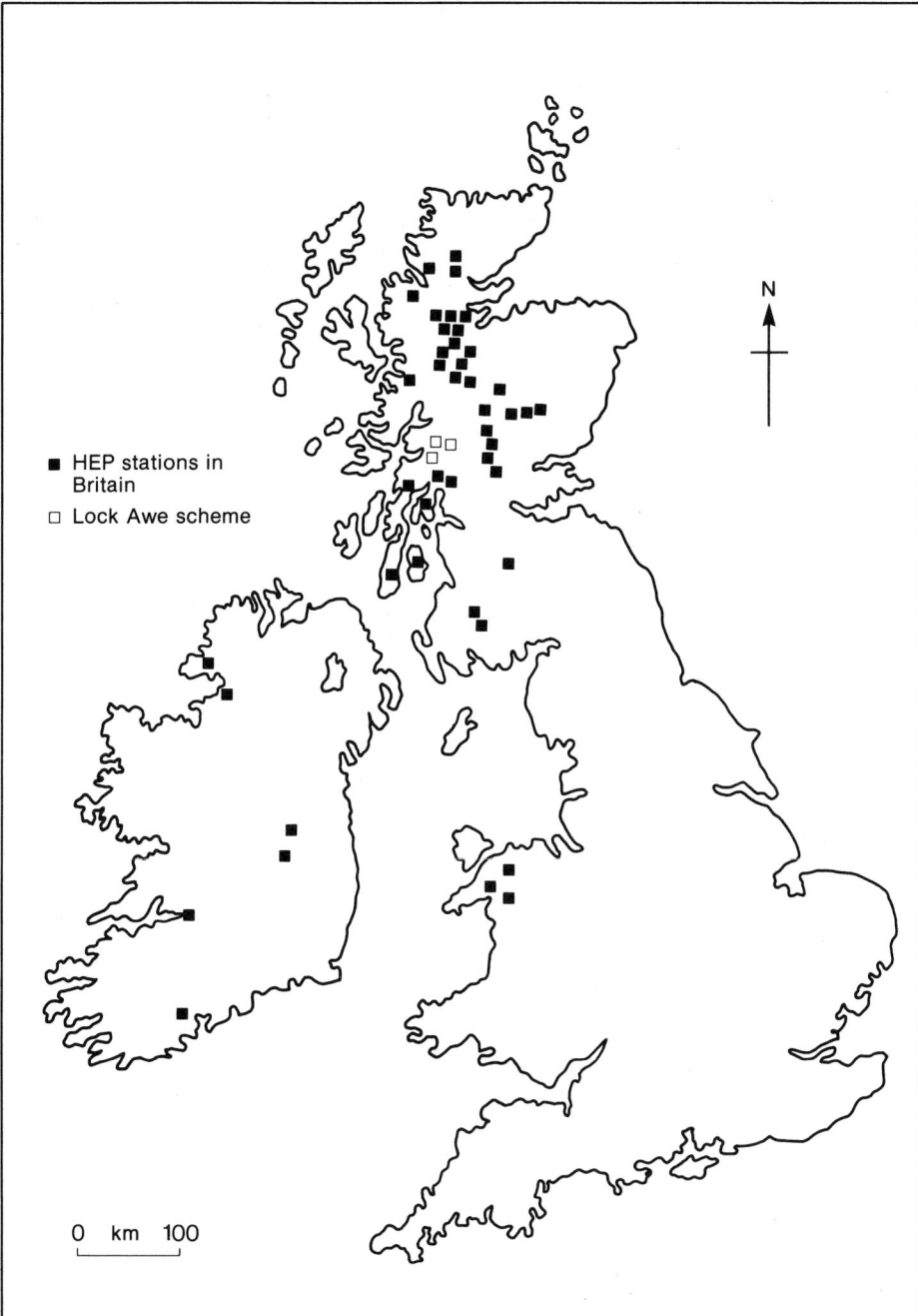

Figure 1.45. *The locations of HEP schemes in the British Isles*

Questions

1 Describe the main areas where HEP is produced in Britain.

2 Explain why HEP stations are located in these areas. (Look at the relief and climate maps in your atlas.)

3 Find East Anglia in your atlas. Explain why there are no HEP stations in that area.

The main HEP stations are shown in Figure 1.45. Sites for HEP stations need:
(a) A regular supply of water. This means they have to be built in areas which have a high rainfall throughout the year.
(b) A narrow valley where a dam can be built. A wide valley would mean a large dam which would be very expensive.
(c) Firm rocks, so that a dam can be built which will not move.
(d) A large head of water, to produce a fast flow of water to the turbines below. The volume of water is more important than the steepness of the slope. Water is stored behind dams to supply a good head of water. If reservoirs need more water, aqueducts and tunnels can be built to divert the water from streams in a different catchment area. A catchment area is the area in which are found all the streams that provide the water for a particular river.

○ THE LOCH AWE SCHEME

Loch Awe is located in the Highlands of Scotland, about 80 km north-west of Glasgow. The site has several advantages for HEP:
(a) There is a high rainfall throughout the year, over 1500 mm per annum.
(b) The area is mountainous, rising to over 1100 metres at one point. The valleys are suitable for dams.
(c) There are many streams that can provide a supply of water.

The scheme was opened in 1963, and includes three power stations. Each has its own supply of water. The two smaller stations, Nant and Inverawe, began producing electricity in 1963. Cruachan started to produce power two years later.

Nant and Inverawe are ordinary HEP stations. They use a strong flow of water to produce electricity all the time. Inverawe is supplied with water from Loch Awe. A dam, or barrage, stops the original course of the water, and diverts it down a tunnel to the power station.

Cruachan (see Figures 1.46 and 1.47) is a much larger power station than Nant or Inverawe. It also works in a different way. It is reversible. During peak times in the day, the station produces electricity like other HEP stations. It generates power using a flow of water from the reservoir to turn the turbine. But at night Cruachan stops producing power. It uses electricity from other power stations in Scotland to pump water back up to the reservoir. Cruachan is then ready to produce power at peak times again the following day.

Cruachan is an example of a pumped storage scheme. (Water is pumped back to be stored for the next day.) Other examples of pumped storage schemes include Foyers in Scotland, and Ffestiniog and Dinorwic in Wales.

Figure 1.48. *Calder Hall, Britain's first nuclear power station*

○ NUCLEAR POWER

Nuclear power is the newest way of making electricity. The first nuclear power station was built at Calder Hall in the Lake District in 1956. Nuclear power now provides about 13% of Britain's electricity needs.

Nuclear power stations use uranium to make electricity. Uranium ore is a radioactive mineral imported from Africa. One tonne of uranium can produce the same amount of power as 20 000 tonnes of coal.

The uranium is used in rods, each about a metre long. When the rods get hot, the heat turns water into steam. The steam turns turbines and can then generate electricity (see Figure 1.49).

Only about 2% of the uranium fed into nuclear power stations is actually used to make electricity. Most of the rest is waste. But some does break down to form small amounts of plutonium. This is a by-product of uranium.

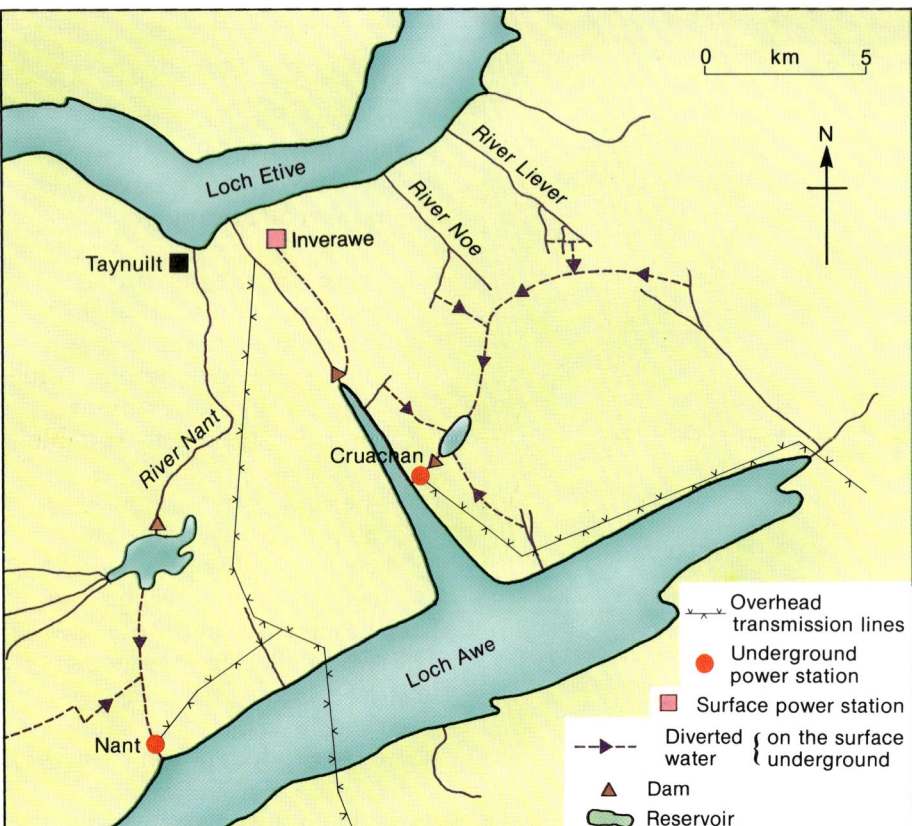

Figure 1.46. *The Loch Awe scheme*
Loch Awe is 40 metres above sea level, while Loch Etive is at sea level.

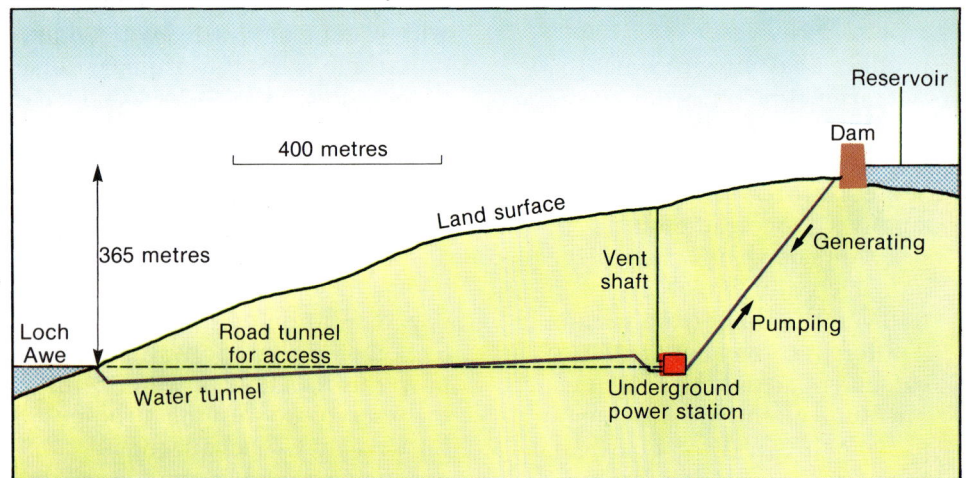

Figure 1.47. *Section through the Cruachan Scheme*

Questions

1 Study the map of Loch Awe in Figure 1.46. Describe how the Nant station receives its water.

2 Describe the other uses that can be made of these lakes.

3 What are the arguments for and against having more nuclear power stations? Collect information from different sources.

Resources · ENERGY

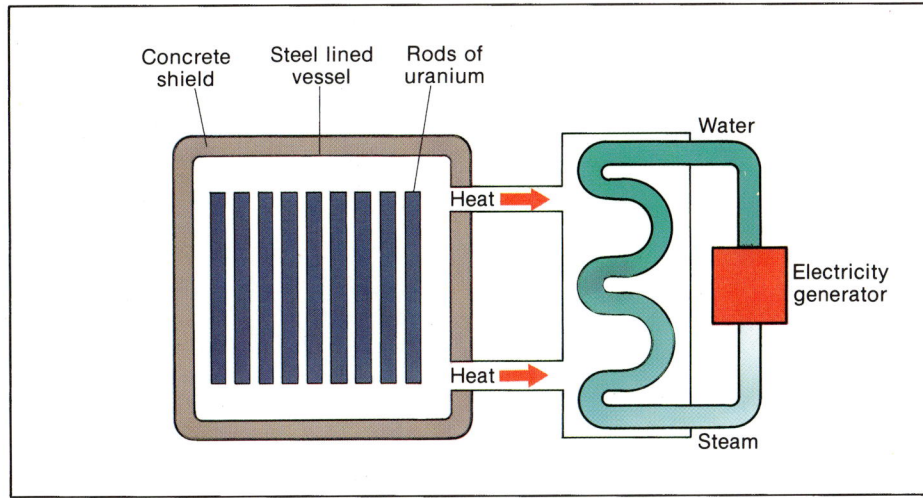

Figure 1.49. *Simplified diagram of a nuclear power station*

Plutonium is another radioactive fuel. It can be used to produce very large amounts of electricity when fed into fast reactors. Fast reactors are a newer type of nuclear power station. The first was opened at Dounreay in the far north of Scotland.

Although nuclear power is cheaper to produce than other forms of electricity, not many nuclear power stations have been built. One of the reasons for this is fear about safety.

Radioactive material is very dangerous. Great care has to be taken to prevent any leaks which could harm people and affect the whole environment. Consequently, uranium and plutonium rods are stored in large steel containers surrounded by a thick concrete shield. Strict precautions are also taken over the disposal of nuclear waste. It is sealed in steel containers. These are either buried deep into rocks or taken out to sea and sunk to a great depth.

Despite such precautions, many people are still concerned that leaks might occur. Some critics also feel that terrorists could hijack supplies of uranium and make atomic bombs. It is difficult to make bombs so this is actually very unlikely. However, the Government has to consider all the facts about safety before allowing more nuclear power stations.

Nuclear power stations need large amounts of water for cooling. All but one are located on a coast. The Trawsfynydd station in Wales is the exception. It gets its water from a large lake nearby.

Originally, all nuclear power stations were built away from large centres of population. But as they seemed fairly safe, a new one was opened close to a heavily populated area. It is situated at Hartlepool in north-east England. All nuclear power stations have to be built on very firm land because of their great weight.

Let us now consider one of these power stations in more detail.

Figure 1.50. *Nuclear power stations*

Figure 1.51. *Dungeness B was the first commercial nuclear power station in the world designed to use advanced gas-cooled reactors.*

Figure 1.53. *Location of Dungeness power station.*

○ DUNGENESS

At first sight, Dungeness may seem a surprising place to find a power station. The area is very flat and low lying. The land is a mixture of sand and gravel, deposited by the sea. The surrounding countryside is mainly grazing land for sheep and cattle.

Figure 1.53 shows that Dungeness is situated on Romney Marsh in an area where the land juts out into the sea. It is actually an ideal site for a nuclear power station.

Dungeness has several advantages as a nuclear power station site:
(a) It is located on the coast and has a constant supply of water for cooling purposes.
(b) It is well away from built up areas, yet close enough to London to have a ready market for its electricity.
(c) The land is firm enough to support the great weight of a nuclear power station.

There are actually two power stations at Dungeness. Dungeness A station was built in 1965. Dungeness B was completed in the 1980s and is of an improved design. It can produce three times as much electricity as the earlier type. As Dungeness B has been able to use the same power lines as A, this has saved money and helped to protect the environment. The countryside in many other places has been spoiled by miles of overhead transmission lines and pylons.

The site at Dungeness is of much interest to naturalists. There is a bird observatory and rare insect colonies as well.

Figure 1.52 *Dungeness B power station near Lydd in Kent*

Resources · ENERGY

Questions

1 Describe the siting of the nuclear power stations in Britain. What are the most important factors which affect their location?

2 With reference to Figure 1.50, choose the best possible site(s) for building a new nuclear power station. Give reasons for your answer.

3 There is much argument at present about whether more nuclear power stations should be built. At a public enquiry, you are to decide whether a nuclear power station should be built, or not. A public enquiry is a public meeting where people speak for or against a major issue. The arguments are heard by someone who decides what should be done. What sort of arguments would you expect to hear
(a) in favour of building the station;
(b) against building the station?

○ COAL-FIRED POWER STATIONS

Just over half of England's power stations use coal to produce electricty. Coal has the advantage of being readily available in this country. There are large supplies underground and it is unlikely that coal will run out in the near future. Because coal is bulky to transport, most of these power stations are found on or near coalfields. In 1985, 87.4 million tonnes of coal were used to produce electricity.

Coal is taken to power stations by rail. When it arrives it is pulverised into dust. The dust is mixed with air and blown into the boilers. A modern power station can consume up to 20 000 tonnes of coal a day. The steam from the boilers turns the turbine blades and this generates electricity. When the steam has turned the turbines, it is changed back to water. This is done in large cooling towers.

Large amounts of water are needed in the power stations so most are sited alongside rivers. There are a number of coal-fired power stations along the River Trent.

Two coal-fired power stations can be found at Rugeley, in the West Midlands (see Figure 1.54). The first, Rugeley A, was opened in 1963. The second, Rugeley B, was opened ten years later.

The large supplies of water needed for the cooling process can be obtained easily from the river. However, water is also stored and reused. The water has to be very pure. It is cheaper to store and reuse it than to purify dirty supplies taken from the river.

The Rugeley site is on a coalfield. A coal mine was built in the early 1960s to supply the power station with energy. The two power stations use up to 18 000 tonnes of coal a day. The Lea Hall pit produces over one million tonnes a year, but this is not enough to provide all the coal needed. Coal is brought from other mines in Staffordshire and the East Midlands. The easiest way of transporting it is by rail. The main railway line passes nearby and trains bring large quantities of coal each day to the power stations.

The site was originally chosen because the land was cheap. Being near a river, the land was too damp to grow crops. It was also unsuitable for any other sort of development, such as housing.

The demand for power comes from the nearby industrial area of the West Midlands. The power produced by A and B stations can supply the energy needed for a city the size of Birmingham.

Although the stations only need a staff of 800, the coal mines nearby employ another 1000 people.

One valuable by-product of a power station is the ash. At Rugeley it is sold to a local factory, where it is made into breeze blocks for the building industry.

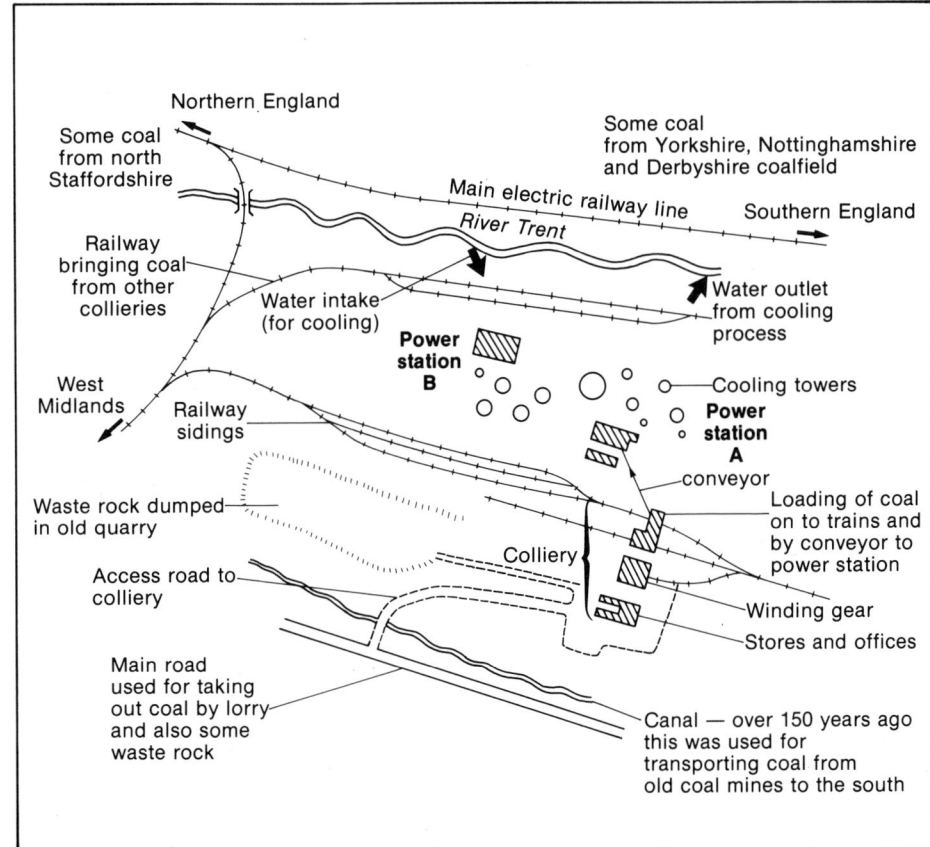

Figure 1.54. *The power stations at Rugeley on the Trent*

The British Isles: Themes and Case Studies

Figure 1.55. *High Marnham coal-fired power station on the River Trent, Nottinghamshire*

○ AN ENERGY PROBLEM?

Over the years, our energy needs have been increasing. It is expected that we shall need even more power in the 1990s and into the next century. Where will we find this energy?

Let us have another look at the types of power already discussed, and see what changes are likely in the future.

Coal has been Britain's main source of energy for about 160 years. We still have large stocks of coal underground. Large deposits are known to exist in several areas of England, particularly the Midlands and East Yorkshire. At the present rate of mining, we should have enough coal for at least 300 more years.

The future of natural gas is more uncertain. Unless new sources are discovered soon, natural gas will run out by the year 2000.

Although the natural gas seems to be running out, there is still plenty of oil under the North Sea and around the other coasts. Oil should supply us with energy well into the next century. New discoveries have been made both on land and in the sea-bed. For example, in southern and central England test drillings have been made for both gas and oil.

Nuclear fuel is considered by many people to be the 'fuel of the future'. There are plans to almost double the amount of electricity produced by nuclear fuel over the next ten years. Nuclear power could solve most of our energy problems. But there is concern about safety. Nuclear power stations may not be developed so much because of these worries.

Hydroelectric power is of limited use in Britain. There are very few sites which have not been developed already. A few more pumped storage schemes are being built to produce electricity at peak times. But HEP will not be able to provide us with much extra energy in the future.

○ WHAT ALTERNATIVES ARE THERE?

During the past ten years, there has been a great deal of research into other ways of providing more energy. It seems possible that coal can be turned into gas and oil quite cheaply.

The action of waves and tides in the sea could provide another new form of energy. For example, tests have shown that waves can produce electricity by making plastic floats rock (see Figure 1.57). But it would take a very long line of floats stretched across a wide area of sea to make much power.

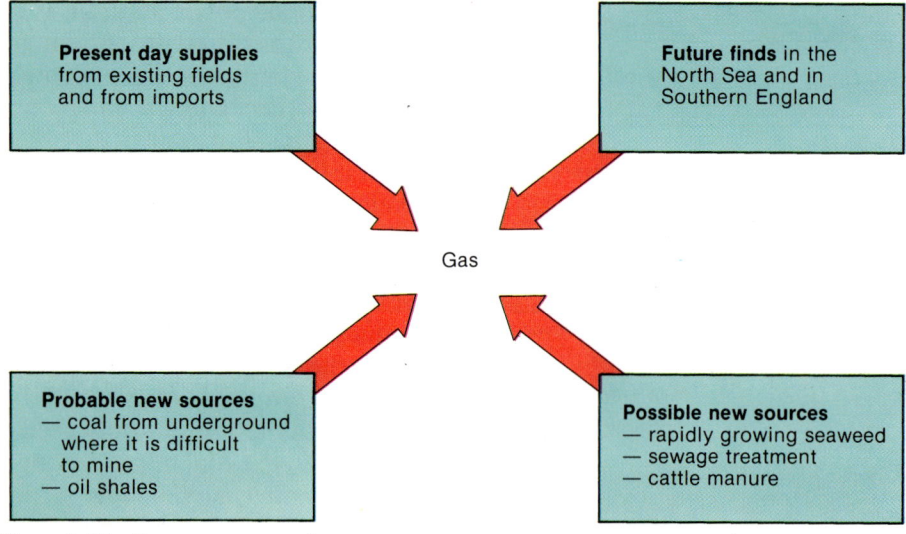

Figure 1.56. *Energy consumption*
Britain's gas supply from the North Sea is expected to be used up in the near future. Much money has been spent building pipelines – so if other sources of gas can be found or made, they can use the same pipes. The diagram shows where we could be getting our gas from in 30 years' time.

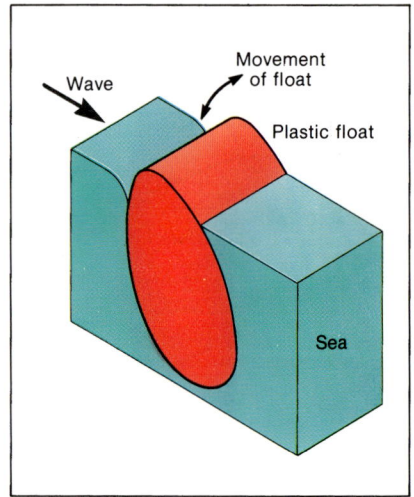

Figure 1.57. *Wave power*
The movement of waves makes the float move up and down. With very large numbers of floats, electricity could be produced in quantity. What might be some of the difficulties of this method?

Solar power (from the sun) is another possibility which is being considered. Solar power is being used widely already in some countries. Britain's first house using solar panels for space and water heating was built in 1975.

The most hopeful discovery may be hot water, found naturally underground. This can produce energy. Water at a depth of 2 km below Dorset has a temperature of 85°C. Another area of hot water has been found 3 km below the New Forest in Hampshire. These finds suggest that geothermal power may be possible in the future.

Our energy problems may not be as difficult to solve as originally thought. But there will have to be changes, and money will have to be spent to research new ideas. Meanwhile, we must take care to conserve what energy sources we already have.

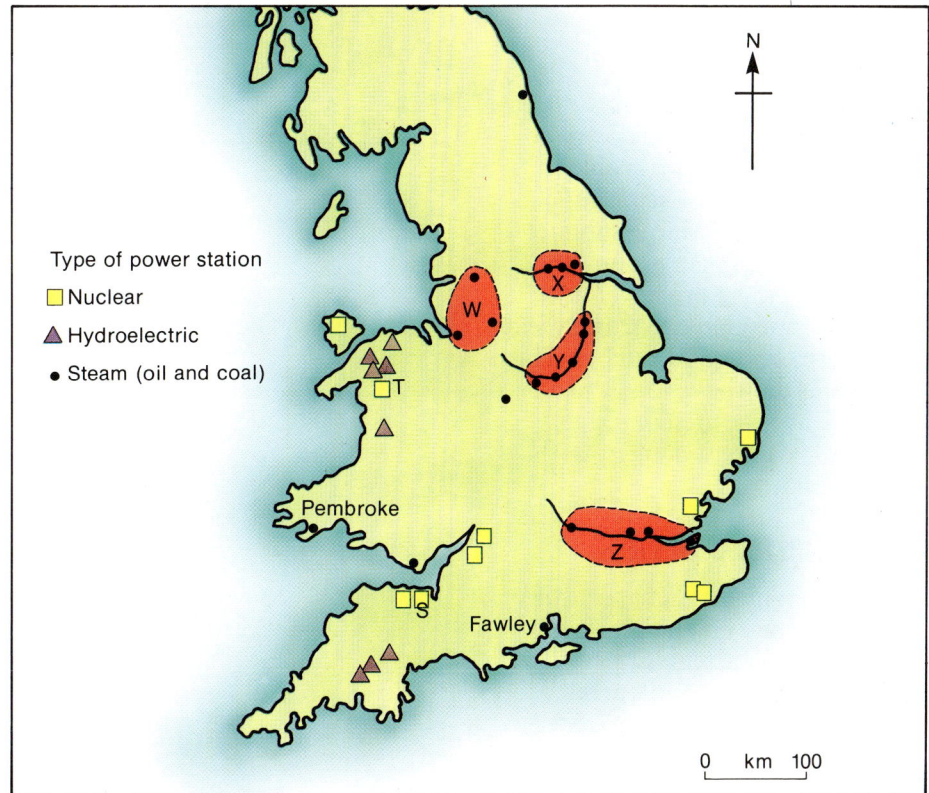

Figure 1.58. *Selected power stations in England and Wales*

Questions

1 (a) (i) Explain why coal-fired power stations are used more often in the British Isles than other types.
  (ii) List two types of coal.
 (b) (i) Name two nuclear power stations near the coast.
  (ii) Explain two reasons for their location.
 (c) (i) What do the initials HEP stand for?
  (ii) Explain three reasons why most of the HEP stations have been built in the highlands of the British Isles.
 (d) (i) What is the area of sea off the British Isles where oil is found?
  (ii) Explain the main advantage that oil has over coal when transporting it to the power stations.
  (South East Regional Examinations Board)

2 Study the map of England and Wales (Figure 1.58), which shows the location of selected power stations.
 (a) (i) On a map of your own, name three of the nuclear power stations.
  (ii) Describe the distribution of nuclear power stations in England and Wales.
  (iii) In what way is the nuclear power station labelled T on the map the odd one out?
  (iv) Describe the advantages of the location of one of the nuclear power stations.
  (v) Explain why two nuclear power stations have been built on sites next to one another at S.
  (vi) State two of the problems facing the future development of nuclear power in Britain.
 (b) The map shows the position of the twenty coal and oil-fired power stations with the highest thermal efficiency in 1985.
  (i) Four areas where several power stations occur have been labelled W, X, Y and Z. Suggest names for any two of these areas.
  (ii) In areas X and Y the stations are coal-fired. Give four reasons why either X or Y is specially suitable for the location of a coal-fired power station.
  (iii) Explain why area Z is an important area for the production of electricity.
  (iv) Fawley and Pembroke are oil-fired power stations. Give three advantages of the site of either Fawley or Pembroke for this type of power station.
 (c) Hydroelectric power stations have been built in Wales.
  (i) Give four reasons why hydroelectricity has been developed in Wales.
  (ii) Explain briefly how a pumped storage hydroelectric power station works.
  (West Yorkshire and Lindsey Regional Examining Board)

Fishing

Being an island, Britain is able to make good use of the sea. Until recently, there were large supplies of many different fish around the coasts. Now there are fewer fish. Some fishermen have to travel great distances to find good catches. This means that they have to spend a long time at sea, often in rough weather. It can be a dangerous occupation.

○ MAIN TYPES OF FISH

There are many different sorts of fish in the sea. Cod and plaice are well known examples. Other varieties, like dab, monk and pollack are not so common. All can be divided into three groups:
Demersal These are fish living near the sea bottom, such as cod and haddock.
Pelagic Fish in this group live near the surface of the sea, and include sprats, mackerel and herring.
Shellfish These live in shallow water near the coasts, and include lobsters, crabs and whelks.

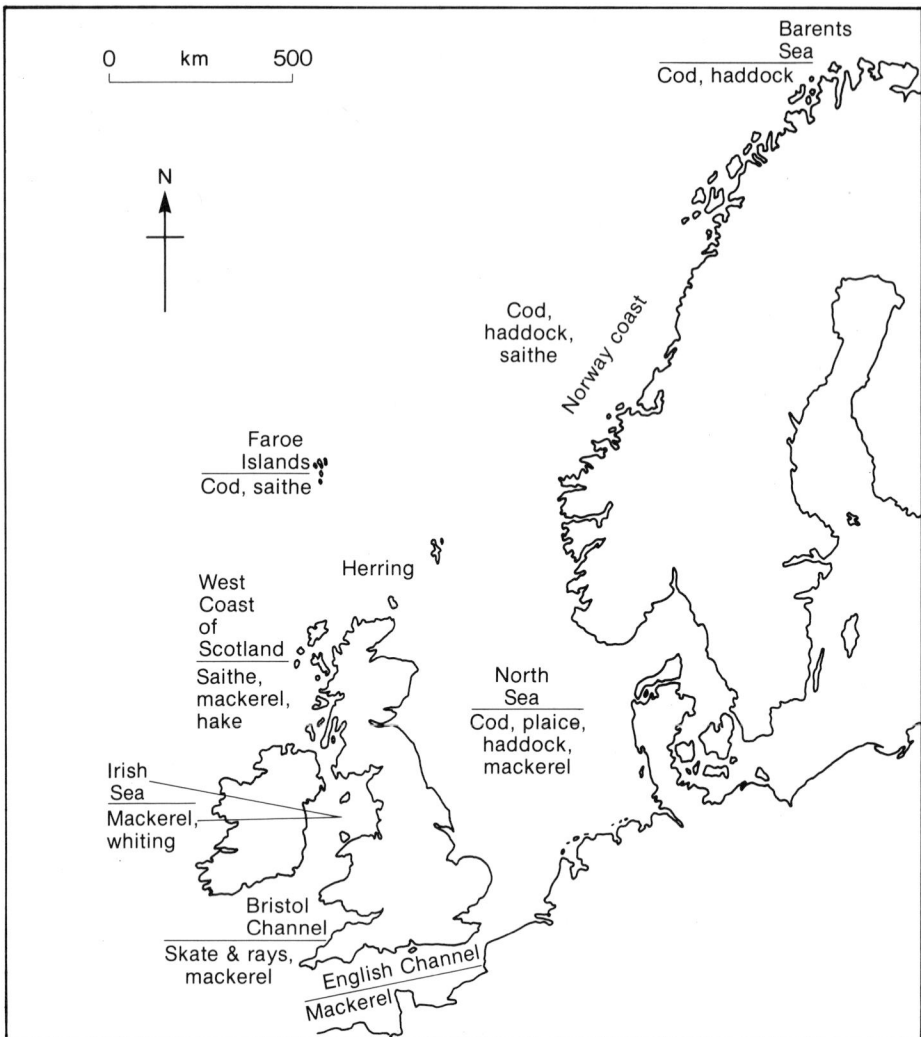

Figure 1.60. *Main fishing grounds*

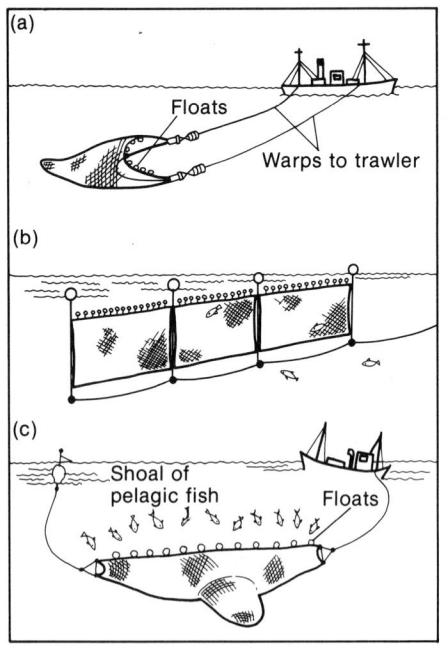

Figure 1.59. *Different nets used for fishing*
(a) *Trawl net*
(b) *Drift net*
(c) *Purse seine net*

○ HOW THEY ARE CAUGHT

Several different methods of catching fish are used around Britain:
Trawling This involves dropping a net in the water and catching shoals of fish as the trawler moves along. The net is tapered. This means that it is wider at the top where the fish enter, and narrower at the other end. It is pulled behind the boat, near to, or actually on the sea floor. Trawlers sometimes work in pairs with a boat at each end of the net. This helps to keep the net open.
Seining This is where a net is used to encircle a shoal of fish. This type of fishing has greatly increased during the last twenty years. One end of the net is fixed to a buoy while the other end remains on the boat. The boat then sails in a large circle around the fish, catching them in the net.
Drift nets These are nets which hang in the water like curtains. The fish are caught in the mesh. Very few of these nets are used today.
Pots and traps These are used to catch shellfish. They are made of wood or metal and plastic, covered with netting. The shellfish are baited into the trap. Up to 60 pots may be placed in a line.

Resources · FISHING

○ MAIN FISHING GROUNDS

Fish can be caught in any of the waters around Britain. Some important fishing grounds are further afield. The main fishing areas (or fishing grounds) are located on Figure 1.60.

The most important area for fish is the North Sea. Almost twice as many fish are caught here as in any other area.

Figure 1.63. *Lowestoft harbour. What area of sea supplies the fish caught by men on these boats?*

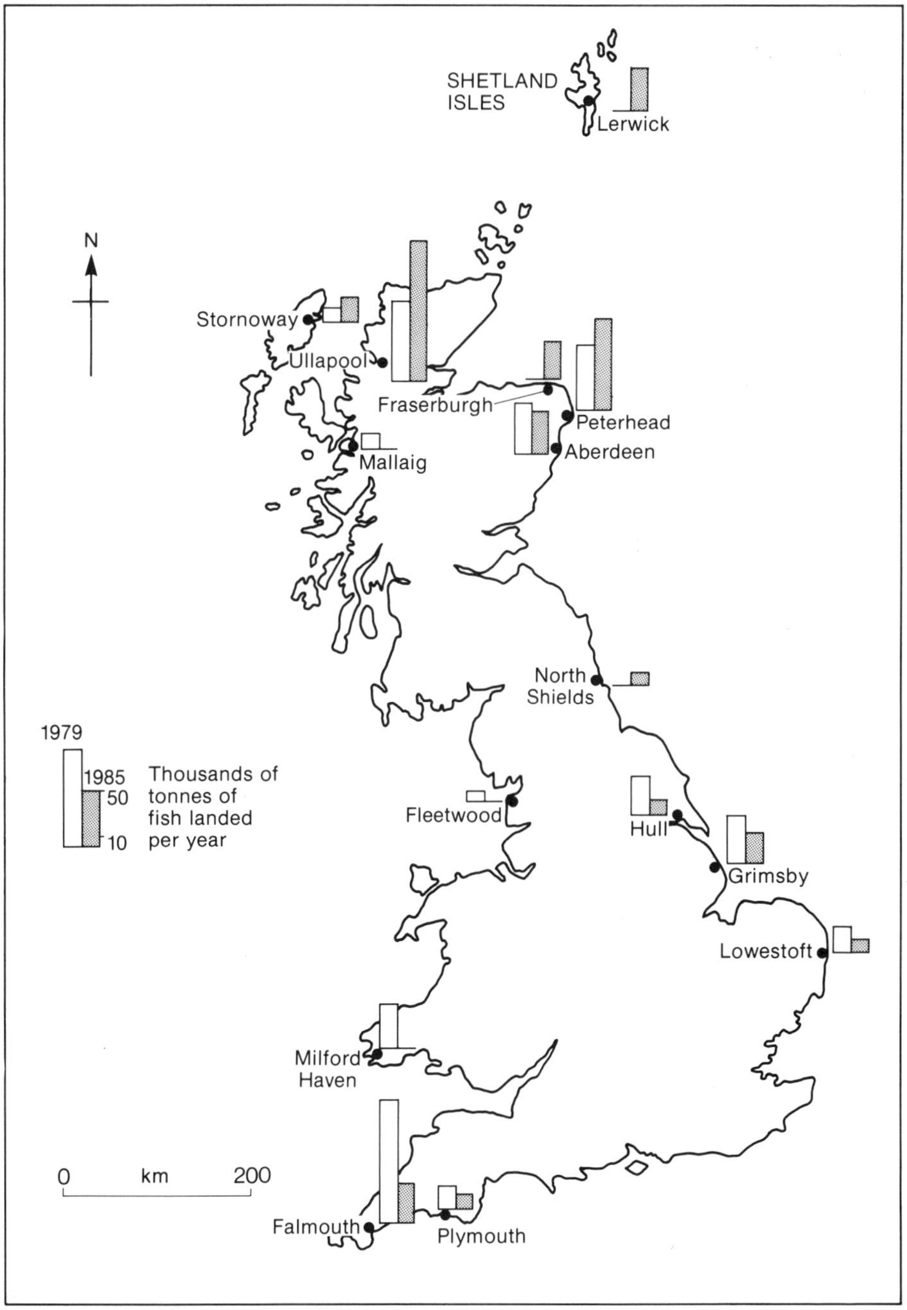

Figure 1.61. *Landings of fish from areas of sea around Britain 1982 (tonnes)*

North Sea	302 000
English Channel	158 000
West coast of England	153 000
Bristol Channel	23 000
Irish Sea	16 000
Barents Sea	6 500
Norway coast	6 300
Faroe Islands	3 100
North West Atlantic	725
Bear Island and Spitzbergen	680
Other areas	16 000

Figure 1.62. *Freshly-caught fish ready for processing.*

Figure 1.64. *Main fishing ports in Britain*

CHANGES IN THE FISHING INDUSTRY

Over recent years, too many fish have been taken out of the seas around our coasts. This overfishing has greatly reduced the number of fish in the waters. For example, in 1980 the mackerel catch was 350 000 tonnes. It had fallen to 175 000 tonnes in 1984. Some fishermen have added to the problem by taking too many young fish which should be left to grow to maturity.

Various attempts have been made to reduce overfishing. The mesh size of nets has been increased. This allows for only larger fish to be caught. Young or small fish can escape. The introduction of quotas for many types of larger fish has meant that only a certain quantity can be caught. New fishing limits have been organised around our coasts. These mean that it is illegal for foreign boats to catch fish close to our shores. All of these measures rely on international cooperation. Countries want to protect their own fishing industry so it has been very difficult to achieve international agreements about overfishing.

There was a rapid fall in the number of fishermen in the industry between 1950 and 1970 (see Figure 1.65). But the number of workers has remained fairly steady since that time, except in the traditional areas. In ports like Grimsby and Aberdeen, the decline has continued. For example, in Grimsby there were 1360 fishermen in 1979, but only 750 in 1985. Aberdeen had 565 working in the fishing industry in 1979, but this figure had declined to 355 in 1984.

Serious unemployment problems exist in places like Grimsby because there is little other industry. Aberdeen is more fortunate. Some jobs are offered by companies working in the North Sea oil and gas areas. However, the decline in fishing has caused some unemployment.

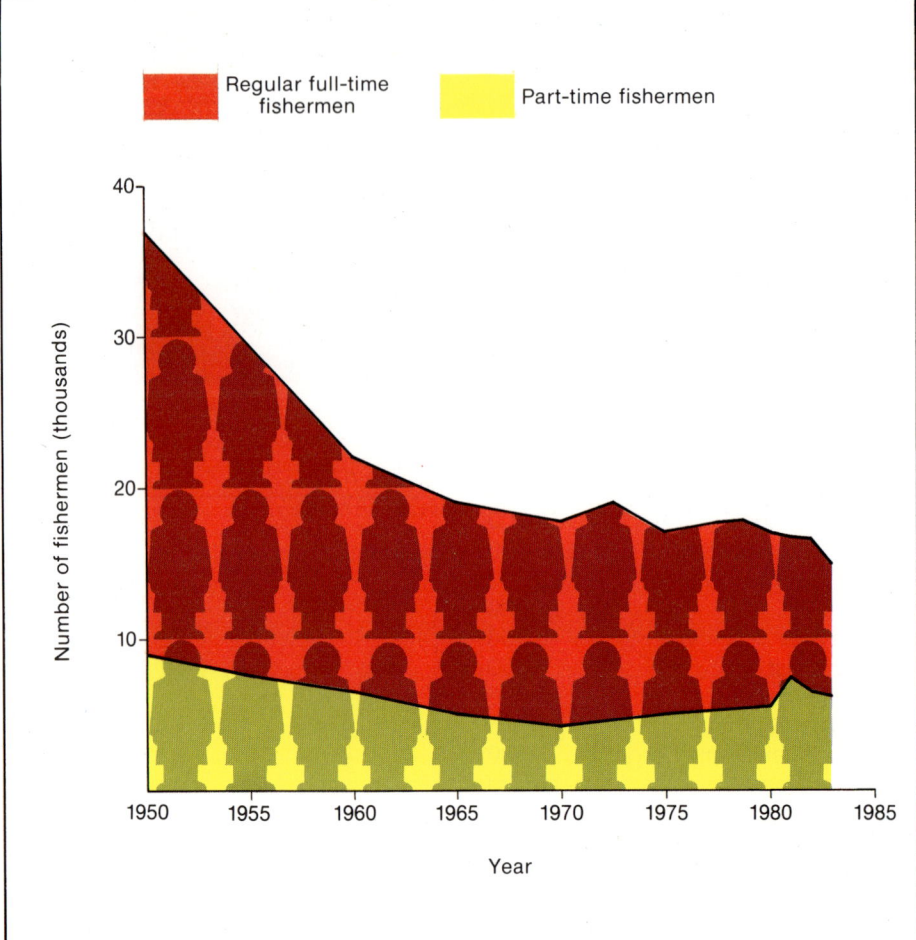

Figure 1.65. *Numbers of fishermen*

Figure 1.66. *Number of North Sea trawlers*

Fishing port	1978	1979	1980	1984
Grimsby	57	41	33	11
Aberdeen	56	42	25	9

As it has become more difficult to find good catches of fish, fewer ships are putting to sea. There has been a particular drop in the number of deep sea ships travelling to distant fishing grounds. Figure 1.66 illustrates this decline.

Fishing boats are generally getting much older. They are very expensive to replace. Some vessels are privately owned by families who cannot afford the high costs involved in buying a new boat. Over a third of all inshore boats are more than 25 years old.

In 1975 over 10% of Britain's fish were caught in the waters around Iceland. But since the Cod War, British fishermen have not been able to fish in Icelandic waters. Figure 1.67 shows the impact of this.

Figure 1.67. *Decline of fish (excluding shellfish) in Icelandic waters (million tonnes)*

	1975	1976	1977
Icelandic waters	94 000	58 000	0
Total fish catch	869 000	933 000	910 000

Resources · FISHING

Questions

1 Describe the main types of fish and explain how they are caught.

2 Study the following table which shows the total fish catch for Britain (including shellfish).

Year	Catch
1971	1 107 000 tonnes
1976	1 063 000 tonnes
1981	859 000 tonnes
1986	730 000 tonnes

What has happened to the total fish catch in Britain?
Suggest reasons for this change.

How can this decline be halted and fish stocks improved?

3 What other changes have taken place in the fishing industry in recent years?

○ FISH FARMING

Overfishing has caused a decrease in the number of fish around Britain's coasts. During recent years many experiments have been carried out to increase fish stocks. Fish are reared and fed in shallow water under carefully controlled conditions. This is called fish farming.

In Britain, trout have been farmed for many years in ponds. Both trout and salmon are being reared now in cages along the sea coasts of western Scotland. Production is increasing, especially for trout. The farming of other types of fish is still limited.

○ USES OF FISH

About one tenth of the fish caught are frozen at sea. This takes place on board large modern, well-equipped vessels with freezing facilities. The remainder of the catch is brought ashore fresh. Some is sold ready filleted, but much of it is bought whole. The fish are packed in open wooden boxes with ice to keep them cool. Fish are then distributed to shops and market stalls throughout the country. Until 25 years ago, most of the fish were transported by rail. Today, almost all travels by lorry. About 22% of the fish goes into fish fingers and other frozen fish products. The lesser known fish tend to be processed into fish cakes. The parts of the fish remaining after filleting are used as fish meal, animal food and fertilizers.

Figure 1.68. *Workers processing fish at Bluecrest Foods Ltd on South Humberside. Most of the fresh fish comes from the North Sea. An increasing amount of fish goes into frozen products.*

Forestry

Wood is the most important building material. It is used for windows, doors and supporting the roofs of houses. Different types of timber have different properties. Some types are ideal for making into pulp and paper. Other varieties are more suitable for furniture. Some woods are extra hard wearing or have a particularly attractive grain. Although we have many forests in Britain, we cannot produce enough timber for all our different needs. We have to import large quantities of wood.

Figure 1.69 shows the main areas of forest in Britain. Forests are found in areas where the soils are unsuitable for farming. The New Forest, Thetford Forest and Ashdown Forest are located on sandy soils. Cannock Forest is on sandy and gravelly soils. Forests can be planted where slopes are too steep to be used as farmland, yet where there is sufficient depth of soil for the trees to take root.

Figure 1.70. *Part of Cannock Forest showing mature softwood trees*
Apart from acting as a firebreak, suggest two other reasons why no trees have been planted in the centre of the photograph.

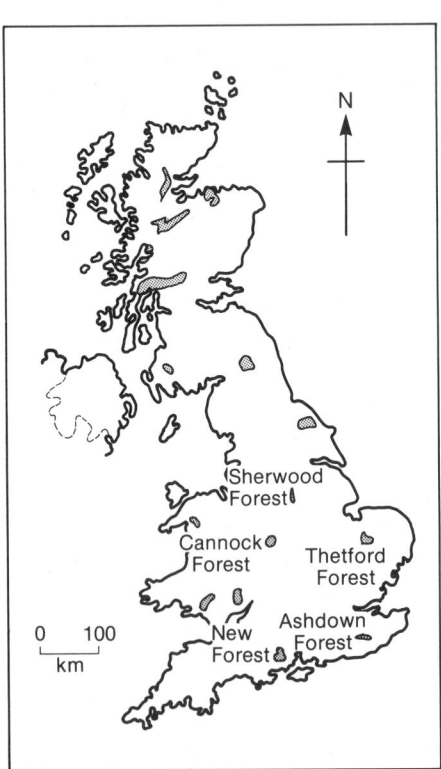

Figure 1.69. *Main forest areas of Britain*

Many centuries ago nearly all of Britain was covered by trees. A few places were too wet or dry for trees to grow and in some places the soil was too thin. However, woodland covered most areas.

As time passed, wood became a very important raw material. Large numbers of trees were chopped down to build houses and the land was cleared to use as farmland. By the early part of this century Britain was running short of trees. As a result, the Forestry Commission was set up in 1919.

One of the Commission's main aims is to provide enough wood for the country in the future. As some trees take over 100 years to reach maturity, this takes a lot of planning. Before deciding what sort of trees to plant, the Commission has to decide how much, and what type, of timber Britain will need in the future.

Different trees produce different kinds of wood. There are two types of tree, hardwoods and softwoods.

Hardwoods
These trees include the oak, elm, ash and beech. They are known as deciduous trees. They shed their old leaves each autumn and grow new ones in the spring. Most of these trees take about 100 years to grow to full maturity, although the oak takes over 120 years.

Softwoods
These trees include the Scots Pine, Douglas Fir and Norway Spruce (the traditional Christmas tree). Most are called coniferous or evergreen trees, since they do not lose their leaves in winter. Each tree takes between 60 and 80 years to reach full maturity. Most are tall and triangular in shape, with many sharp, needle-like leaves, and often cones.

Hardwood or softwood

Softwood trees grow almost anywhere and reach maturity quite quickly. Hardwoods take a long time to grow, and only do well on certain soils. As a result, many more areas have been planted with softwoods than hardwoods. This means that timber from hardwood trees is much more expensive to buy.

The softwoods tend to be used more extensively. They are used for floors, roof supports and window frames in houses. The hardwoods are more resistant and hard wearing. They are generally more expensive and are used for specialised requirements.

Figure 1.72. *Some important uses of timber*

SOFTWOOD	
Norway Spruce	boxes, cable drums, kitchen furniture, pulpwood, woodwool (used in packing)
Douglas Fir	carpentry work, packing cases, roof supports, furniture, telegraph poles, manufacture of fibre board.
Pine	kitchen furniture, boxes, flooring, pitwood (used in coal mining), particle and fibre board
Larch	pitwood, telegraph poles, supports for river banks, boat building
Western and Red Cedar	greenhouses, seed trays, cladding, fencing
HARDWOOD	
Elm	coffin boards, sea defences
Oak	expensive furniture, sea defences, barge building
Ash	furniture, sports equipment
Beech	furniture

Conservation

The woodland provides important habitats for wild life. This is particularly true of the older forests in England. Glades are left for deer to graze. On a small scale, it is important that dead wood is not cleared completely. Dead trees lying on the ground encourage the growth of fungi and insect colonies. Without the insects, the whole web of life in the forest can be upset. The overall appearance of the forest is also important. Large areas of the same kind of tree can be monotonous. Where possible, different species are planted in blocks to form an attractive landscape.

Recreation

Many forests can be visited by the public. Most visitors want to park their cars, find a picnic space with a view, and go for a short walk. But too many visitors can cause problems. Some people light fires, or leave litter, or damage young trees. Several forests now have specially laid out forest walks. These allow visitors to identify the various trees without wandering about too freely.

Leaving space for recreation means planting fewer trees, and so producing less timber for the country. Many conservationists also want more of the countryside left in its natural state.

Questions

1 Make a list of items in your home that are made from wood. Try to suggest what type of wood has been used.

2 Apart from providing timber, what are the other important functions of forests?

3 Do you consider that forests are finite or infinite resources? Give a reason (refer to **Resources** on page 1).

Map reading questions

Look at Map 2 on page 124, and answer the following questions:

1 Describe the extent of the forestry on the map. You should say whether it covers all the lowland or all the highest peaks; whether it is found on the steepest slopes and whether there is an upper level of tree planting.

2 List five ways in which tourists may be helped in the forest. For each example give a grid reference.

3 Why is part of the forest bounded by a thick yellow line? Suggest what this might mean to holiday makers as well as local people.

Figure 1.71. *Trees such as silver birch are grown in the forest*
The area shown here has very stony soil. These trees are growing around the edge of the forest. Conifer trees can be seen in the background.

○ CANNOCK FOREST

One typical forest is Cannock Forest in the West Midlands. It is 30 km north of Birmingham. The forested area covers 2700 ha of land. There are in fact several woods separated from the main area of forest. A number of roads cross the area and much of the land is accessible to the public.

The soils are very acid sands, containing a high amount of gravel. The low rainfall in this area combined with the soils means that the land is very unsuitable for farming.

The forest was one of the first created by the Forestry Commission. Cannock was set up in 1922, three years after the Commission was formed.

The main trees found in the forest are Scots Pine and Corsican Pine. At the moment there are roughly the same number of each. These trees grow well in acid soils. In addition, there are small areas of larch, and a few broad-leaved trees, like oak. They are found in the southern part of the forest where the soils contain a high proportion of clay.

The Corsican Pine takes 50–55 years to mature, while the Scots Pine takes between 60 and 65 years to reach its full size. This means that during the 1980s, many parts of the forest have been cut down for the first time. Replanting is taking place, mainly using Corsican Pine rather than Scots Pine. This is because the Corsican Pine produces a better yield of timber, takes less time to grow and is not so affected by diseases and pests.

Areas to be cut down (or 'harvested') are chosen very carefully. This prevents large areas being cleared at any one time, which would be unattractive and would destroy the habitats of some animals. Almost 40 ha of the forest are being cut down and replanted each year.

The young trees used for replanting begin life as seedlings. After growing for more than a year they are transplanted to the forest. There they are placed in wide, shallow furrows which have been turned over by a plough. Up to 2500 young trees are needed for every hectare which is replanted. During the last few years, experiments have been taking place using pot grown plants. These are transplanted into the soil from their pots. Young trees planted in this way take less time to become established.

Young trees are planted in straight rows, with rides or wide tracts left between each area. This allows foresters' machines and lorries easy access.

The new trees are planted very close together. This stops grass and other plants growing around them. After about six years, the trees are thinned out for the first time.

Figure 1.73. *One of the walks in Cannock Forest*
Families are encouraged to use the carefully planned walks. This photograph shows the start of a 45 minute walk.

Further thinning takes place after twenty years. The remaining trees are then left to reach maturity.

About 14 000 tonnes of timber are obtained from Cannock Forest every year. One third of this, mainly the later thinnings, is used for pit wood. The timber is used for pit props in nearby coal mines. It is collected and transported by an outside contractor. The remaining timber comes from the final cutting of mature trees. Most of this is made into saw logs which are roughly cut lengths of wood. These are delivered to many sawmills throughout the West Midlands. There they are used for such items as fences, sheds and benches. This provides numerous jobs outside forestry but still working with timber.

There are few diseases that affect the trees in the forest. But the 'pine looper' does cause some problems. It is a caterpillar which eats the needles of the Scots Pine. This weakens the tree and leaves it open to attack from other insects, particularly beetles. The best method of controlling this pest is to cut down affected trees.

Animals cause some problems. The deer, for example, tend to rub their antlers against the trees and strip the bark. Providing that they do not cause too much damage, the foresters are not concerned. The herd has started to increase recently to a certain extent, but the numbers are controlled by road deaths. On average, two are killed by cars each week. There are now about 300 deer in the forest.

Strong winds and very dry summers can cause great damage. In 1976 many trees were uprooted or broken by gales. However, people often pose the greatest threat to the forest. Fires that are started deliberately and those caused by carelessly thrown cigarette ends have resulted in thousands of pounds worth of damage. These fires have destroyed large areas of valuable woodland.

The Forestry Commission actively tries to encourage people to use the forests (see Figure 1.74).

Figure 1.74. *Visitors to Cannock Forest*

This is particularly true of Cannock Chase because it is the closest area of woodland to Birmingham and the northern part of the West Midlands.

Landscaping has been carried out in many areas to make the forest look more attractive. This involves planting different varieties of trees in small clusters, particularly where rides cross each other. Support has been given to a number of leisure activities. These include orienteering, fishing in small lakes, walking, and pony trekking.

In this forest, there are four supervisors and a head forester. A further 24 employees are skilled forestry workers. In addition, about six other workers are employed by outside firms. Although the forest is large, it does not require many people to look after it.

Questions

1 From the information on these pages, describe the jobs that foresters do throughout the year.

2 Explain why the area of Cannock Forest is well suited for forestry rather than farming.

3 What other advantages and benefits does forestry provide in addition to the growing of trees?

4 Many visitors come to Cannock Forest. By far the largest number travel by car (98%). It is a popular area for cycling clubs to visit from nearby towns. In addition, a few use the bus service, especially at weekends.

In one survey from a week in May, the number of visitors was 80 on Monday, 90 on Tuesday, 128 on Wednesday, 40 on Thursday, 110 on Friday, 320 on Saturday and 545 on Sunday. Draw a bar graph of these figures and describe what the results show. Is it what you would expect?

Map reading questions

1 Look at Map 3 on page 125.
(a) What can be found at 043128? How high is it?
(b) What do the letters CH at 028137 stand for?
(c) What type of routeway is Marquis's Drive at 030148?

2 Using the evidence from Map 3, describe the relief of the area covered by the large area of woodland. Include the height of the land, the highest and lowest parts, and the steepness of the slopes.

3 Describe the different roads and paths that could be used to reach the central parts of Cannock Forest (Map 3).

SECTION 2 · Farming

There are 56 million people living in Britain, and they all have to eat. Some products, like bananas and rice, have to be imported. But much of the food in our weekly shopping baskets comes from farms in this country.

Farming includes raising livestock, growing arable crops and market gardening. Livestock farmers keep animals like dairy cattle, sheep, pigs and poultry. They may also grow some crops as winter fodder for their animals. Wheat and barley are the main arable crops. Some arable farmers also grow a few vegetables. Most vegetables, flowers and salad crops are produced in market gardens.

Figure 2.2. Farming production figures (million tonnes)

Product	1950	1980	1984
Cereals (million tonnes)	15.4	17.3	26.5
Meat (million tonnes)	1.6	2.9	3.2
Milk (thousand million litres)	10.6	16.0	16.4
Sheep (millions)	22.7	30.0	34.8

The number of people living in Britain is gradually increasing but the number of people working on the land is decreasing. This is in spite of the fact that we are trying to produce more and more of our own food. In 1950 4.9% of the working population in Britain were employed in farming. This figure had dropped to 2.6% by 1980 and remained the same in 1984. In 1984, 34.7% of the workforce were in industry and 62.6% were employed in the service sector. In other European countries a rather higher percentage of people work in agriculture. It is 6.2% in West Germany and 8.8% in France.

Though fewer people are now employed in farming, the production figures for many products have actually increased. Farming has become more intense and highly mechanised.

To get the most out of the land, farmers have to know about the climate and their soils. Certain areas are too cold for some types of farming, while others may be too wet or too dry. Farmers have to make certain their soils suit the type of farming they choose. Soils have to be well maintained to produce the highest yields. This can be done by adding fertiliser or manure.

Farmers have to be aware of change. Competition from overseas producers, and changes in the market, mean that farmers have to introduce new methods and different strains of crops. Competition for agricultural land from builders and planners creates other problems for the farming industry.

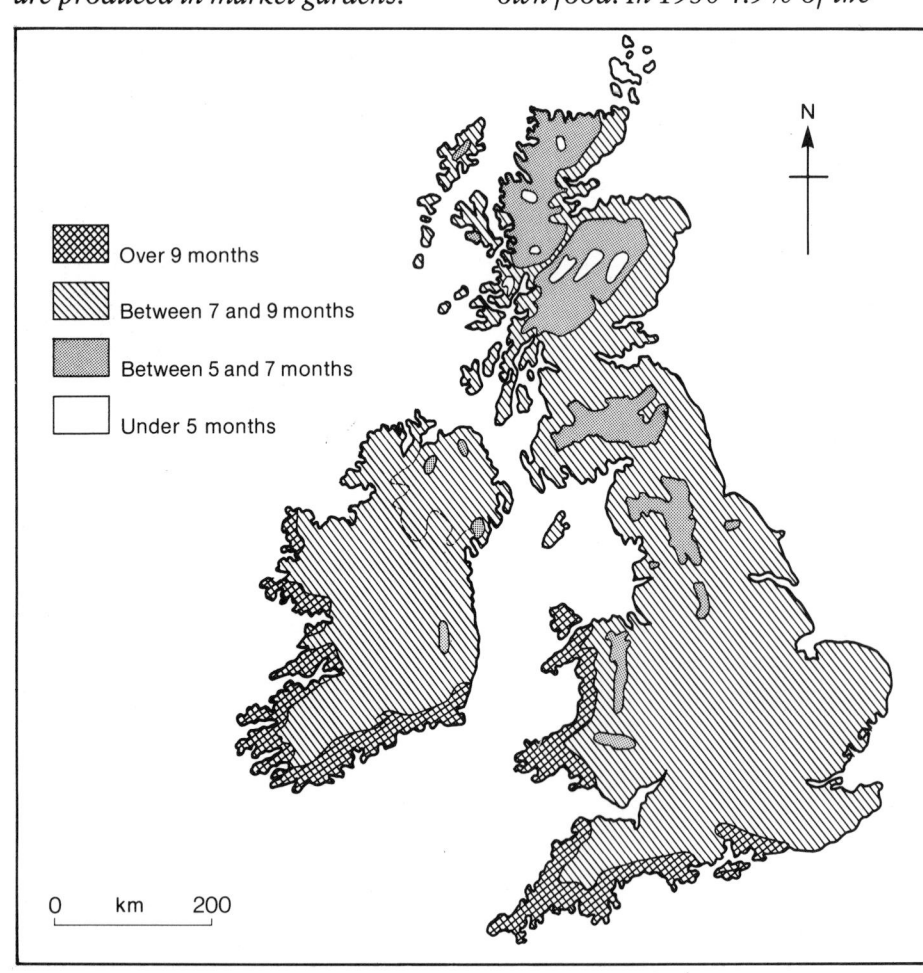

Figure 2.1. *Length of the growing season*
This is the number of months where the temperature is above 6°C. Using an atlas, describe the parts of Britain where (a) the growing season is longest and (b) the growing season is shortest. How do you think this influences what the farmer grows?

Figure 2.3. *The varied farmland*
This aerial view shows the varied nature of British farmland. Describe the view shown in the photograph.

Farmers need to make a profit from the land. Generally, the most expensive land is found around towns and cities. Farming in these areas has to be very intensive. Land is usually cheaper further away from urban centres. Farming here can be more extensive. With more land available, the farmer's costs are less. In particular, transport costs are lower. For example, produce from sheep farms is not sent to market very often. Farmers nearer towns, such as market gardeners, may have to deliver to their customers every day.

Figure 2.5. *Harvesting barley. Why has mechanisation led to fewer and fewer people working in farming?*

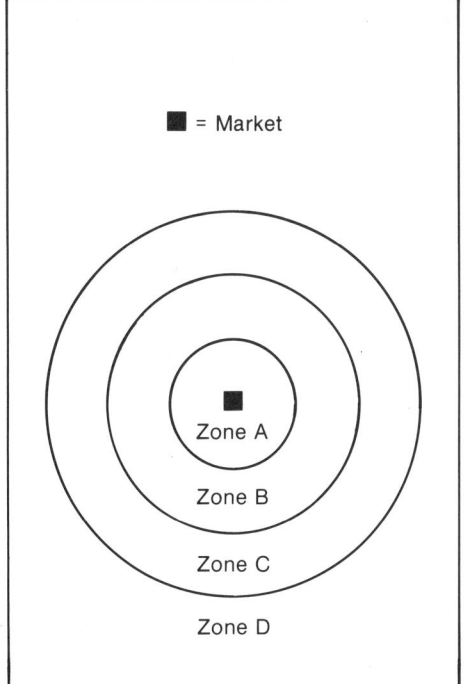

Figure 2.4. *Market zones*

Exercise

After carefully reading through this introduction, look at the information in the table below. It shows the costs and selling prices of four types of farm produce.

1 Copy out the table. Work out the profit made by each type of produce. (Subtract the cost from the price of the produce).

2 Copy out Figure 2.4 and try to decide the best type of farming for each zone: A, B, C and D. Write in the type you choose in place of the letter.

Farm produce	Price of produce per ha (£)	Cost of produce per ha (£)	Total profit (£)
Market gardening	1 200	900	
Dairying	600	400	
Cereals	300	150	
Sheep farming	200	100	

Dairy Farming

We all know that cows produce milk. But our daily pint depends on dairy farmers. They make sure that their herds produce the quantity of milk we want, when we want it. Regular supplies are essential to the dairy industry.

As cows only produce milk when they have calved, most farmers arrange that each cow has a calf every twelve months. Each cow may then produce milk for up to 200 days of the year. Dairy farmers carefully plan when their cows have calves. They can then ensure a regular supply of milk from their herds all the year round. The young bull calves are usually sent to market for fattening and then slaughter. Farmers keep the young cows (heifers) so that they can be added to the herd when they are old enough to produce calves and milk.

Dairy cows need plenty of grassland on which to graze. Quite a heavy rainfall is needed to keep the grass in a lush condition. Areas with over 700 mm of rain a year produce the best grass. The type of soil is important too. Clay soils retain the water to a greater degree than sandy soils, which allow the water to pass through quite freely. The richest dairy farming tends to occur in areas of high rainfall and clay soils.

Farmers depend on grass for more than just summer grazing. Grass also provides the hay and silage which is stored for winter feeding. Some grassland is permanent. This means that it has not been ploughed and specially planted with grass. Year after year it remains as grass. But many farms also use temporary grass. Here the land is ploughed and planted with special grass called ley grass. Such fields may be in use for up to five or six years. Fertilisers often have to be added to keep the grass in good condition. A system of land drains is often added to make certain that waterlogging does not occur.

Figure 2.6. *Dairy cattle*
The cattle shown are Friesians. They are the most popular dairy cows because of their high milk yield. Over 80% of our milk comes from Friesian cows.

During the winter, cows usually live in barns. Most farms have large barns where the hay and silage for winter feeding is also stored. Although they graze outside in the summer, cows are brought indoors to be milked twice a day throughout most of the year. Most farms have modern well-lit milking parlours. Using modern machinery, one person can milk over 100 cows an hour.

Some dairy farmers let their cows graze wherever they like in the fields in summer. This can lead to the corners of fields being overgrazed and other parts not being grazed at all. As a result, many farmers now control the way their animals graze by putting up electric fences. Only one strip of grass can be grazed at any one time. The fence can be moved after each strip has been grazed.

As well as grass, most farmers also grow other crops like barley, oats, sugar beet and kale. These are used as winter food for the animals.

The main dairying areas are in the west of Britain. South-west England, Cheshire and the western area of central and southern Scotland are the most important. In addition, there has always been a concentration of dairy farming around large towns and cities. This is because fresh milk should be delivered to customers as quickly as possible. With faster transport this is now not quite so important. Any surplus milk is made into butter, cheese and yoghurt.

Farming · DAIRY FARMING

○ THE COMMON AGRICULTURAL POLICY (CAP)

Britain joined the European Economic Community (EEC) in the early 1970s. The EEC's Common Agricultural Policy (CAP) guarantees good prices for farmers, encouraging over-production. The biggest surpluses have been in milk and cereals. Foods such as butter have been stockpiled in 'mountains'. Some foods have been destroyed.

The high cost of the CAP has meant shoppers paying higher prices. Farmers did well in the 1970s as prices were supported. Many farmers borrowed heavily to buy machines and other aids. From the mid-1980s, production was being limited to reduce surpluses. In 1984 a milk quota began which meant less earnings for UK dairy farmers. Many UK farmers have been in trouble in the 1980s because of their debts and less certain earnings.

Figure 2.7. *Number of farms producing milk*

	England and Wales	United Kingdom
1955	143 000	175 000
1960	123 000	152 000
1965	100 000	125 000
1970	80 000	101 000
1975	60 000	77 000
1980	43 000	56 000
1984	39 000	51 000

Figure 2.8. *Average size of the milk producing farms (hectares)*

	1972	1978	1984
England and Wales	52	65	67
Scotland	75	84	90
Northern Ireland	23	30	36

Questions

1 Draw a graph to illustrate the information in Figure 2.7. What trend has taken place? Has it been a steady trend? In 1980, what proportion of total milk producers in the United Kingdom were in England and Wales?

2 What does Figure 2.8 show about the size of milk farms in the various parts of Britain?

3 On an outline map of Britain, mark in Cornwall, Somerset and Cheshire. Using an atlas, shade in lightly the areas of grassland (permanent pasture).

4 Write an account to explain why dairying is mainly in the western half of Britain. (Think of the climate, relief and markets.)

5 From the information provided in this section, write an account of the changes that have taken place in dairying in Britain.

Figure 2.9. *Park Farm*
Note the long lush grass in front of the farm.

○ PARK FARM

Park Farm is in Cheshire, a very important area for dairying. The farm is just over 105 ha in size, which is quite large for this area. Park Farm is linked to nearby villages by a road which connects with six other farms.

As can be seen from the plan in Figure 2.10, a stream flows through the farmland. The stream is not especially deep or wide, but in very wet weather some of the grassland downstream tends to be flooded. Much of the land belonging to Park Farm lies on the sides of a low valley. This is between 60 and 80 metres above sea level. Two areas have remained as woodland – one near the meeting of the streams, and the other near the road. The soils are almost all clay. The clay helps to retain moisture in the ground, and keep the grassland in good condition.

The rest of the farm is divided into ten fields. There used to be many smaller fields, particularly in the northern part of the farm. This area is now two large fields, with 20 ha each. The hedges have been removed and electric fences control the grazing of cows.

There are 280 cattle producing milk on the farm. A further 40 heifers are in calf, and there is a pedigree bull. All the cattle are Friesians. The farmer has tried other varieties of cattle but he has found that Friesians are best suited to this area. In 1968, the whole herd had to be destroyed due to an outbreak of foot-and-mouth disease.

The cattle are kept indoors from early November to the end of March. During this time they are fed on hay cut from fields during the summer (silage) and oats. For the rest of the year the cows are left to graze.

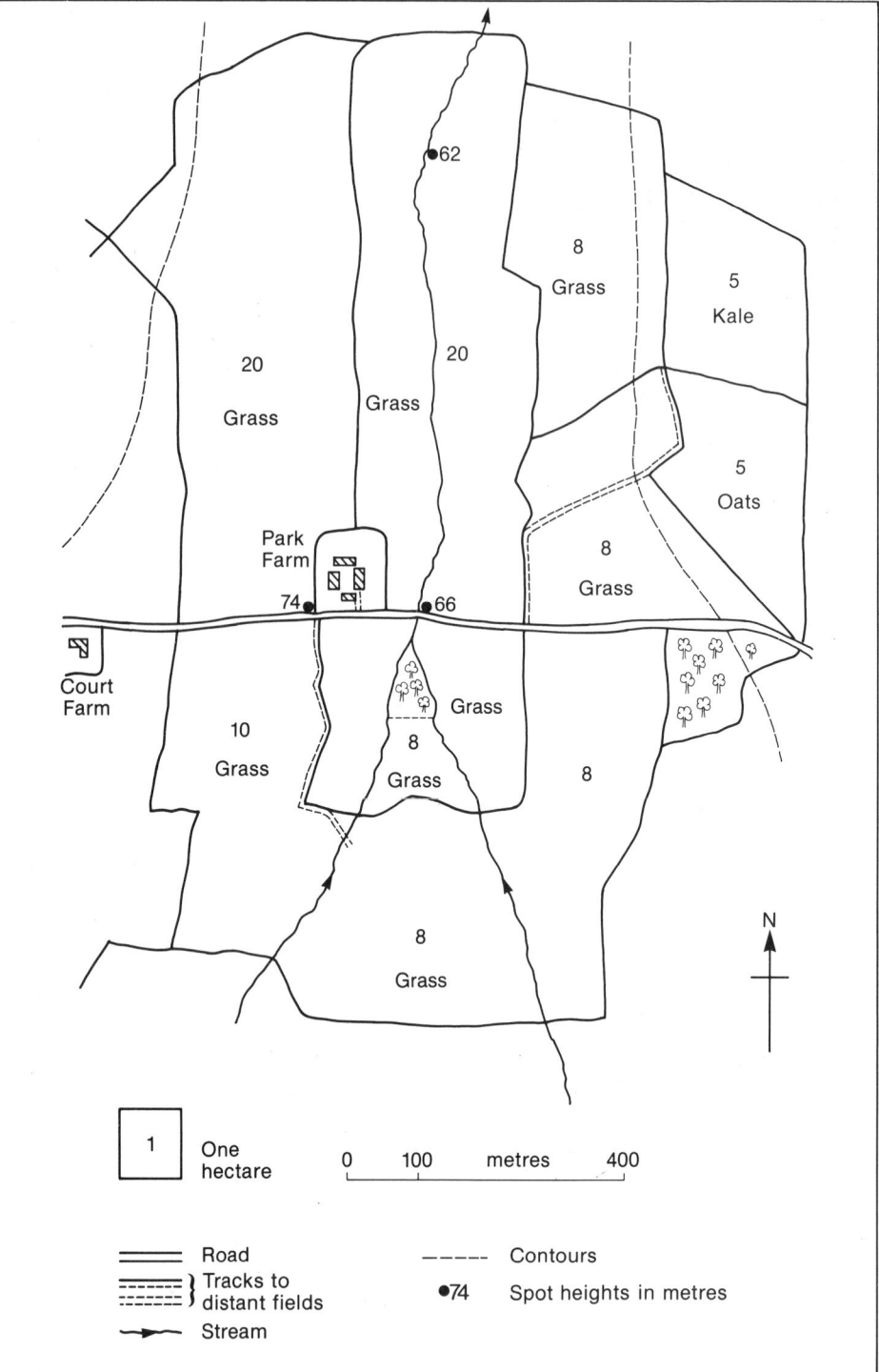

Figure 2.10. *Park Farm*

Milking takes place twice a day, generally at 6.00 am and between 3.00 pm and 4.00 pm. The farmer prefers to keep the cattle near to the farm as it is easier to bring them in for milking.

The milk yield is usually slightly higher than the average for the rest of the country. In 1981 it was 5305 litres per cow. The cows are milked in large parlours, using the most up-to-date methods. The milk is collected each day by tanker and taken to the local dairy 30 km away. Several years ago, some of the milk was taken to be turned into Farmhouse Cheese. Now this cheese is made from milk taken only from a few selected farms.

Farming · DAIRY FARMING

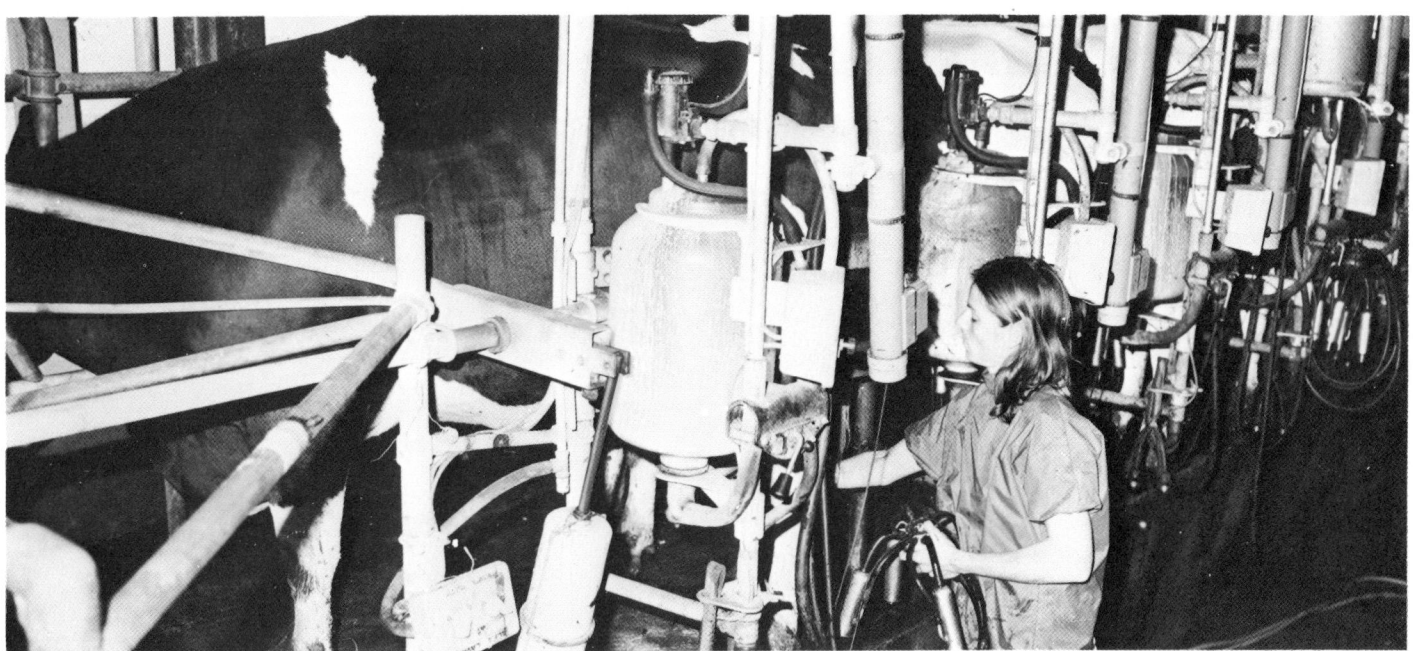

Figure 2.11. *A milking parlour*
The milk passes through a milk recording jar, before going to a storage vat. There is a refrigerated storage tank where the milk awaits collection.

There are two fields used to produce winter feed. These are marked on the map (Figure 2.10). Each is 5 ha. At varying times these fields have grown oats, barley, kale and turnips. Of the other fields, the 20 ha and the 8 ha fields by the stream are permanent grass. Most years they are too wet and heavy to be ploughed. The remaining fields have been ploughed and sown with special grass seed. These are called Ley Fields. They need re-seeding with grass about every five to seven years. Fertiliser is added depending on the soil conditions. Not all the fields are used for grazing. Some are left to be cut for hay and silage, which is stored in the barn for winter feeding.

The farmer has kept two areas of woodland. This is largely for his own use, as he allows shooting to take place on this part of his land. But conservation is another important reason why he has preserved the woodland. Over recent years, thousands of trees and hedges have been felled and the land ploughed up to make more farmland throughout the country. The woodland here is a small attempt to preserve the countryside in this area.

Figure 2.12. *Milk tanker*
Once a day the milk is collected from the storage vat by tanker. It normally happens in the morning after the first milking. The tanker driver tests the milk before it goes into the tanker.

Questions

1 The numbers in the fields in Figure 2.10 give their sizes in hectares. Only the fields with numbers belong to the farm.
(a) What is the total area of the fields in hectares? Count up all the field sizes. The actual size of the farm is just over 105 ha. Explain the difference. Estimate the size of the area of woodland.
(b) Draw a graph to show the amounts of land used for grassland, arable land, and woodland.
(c) Write an account of what you think a typical day for the farmer might be. Consider other jobs not mentioned that the farmer has to do.

2 Which fields are likely to be used for grazing and which are likely to be used for hay. Think about the need to collect the cattle twice a day for milking.

Arable Farming

Arable farming is the growing of crops. There may be a few animals on an arable farm, but the main work is to produce crops. These crops include wheat, barley, oats, potatoes and sugar beet. Arable crops can be divided into two main groups:
(a) those that are to be sold (called cash crops);
(b) those that are grown to feed the animals on the farm (called fodder crops).

Sometimes a farmer might grow a crop like barley to feed his animals one year. The next year he might decide to sell the barley. The first year he would be growing a fodder crop, and the second year his barley would be a cash crop.

Of the cash crops, wheat, potatoes, barley and sugar beet are the most important. Crops such as wheat and barley are called cereal crops. To grow these crops, the soil has to be ploughed before the seed can be grown. This allows the frost to break up big lumps of soil into a fine tilth. The seed is planted in the spring and harvested in late summer. These are called spring crops. Some cereal crops are sown in autumn so that they can spend winter in the ground. They can then be harvested in mid-summer. These are called winter crops.

Several factors affect arable farming.

Relief and soils
The land needs to be flat or gently undulating (sloping). This allows machinery to be used easily. Much of south-east and eastern Britain is below 200 metres above sea level. There are few hills to contend with. The soils need to be rich in humus and retain moisture well. In East Anglia ice covered the area during the Ice Age. In many places a deposit of clay and boulders was left behind. This is called boulder clay. Such soils are well suited to arable crops. The ice only extended as far south as the River Thames. Further south, the soils are varied. On the gentle slopes of the chalk hills there are clay soils lying on top of the chalk. Once it was grassland and animals grazed the slopes. Today, much of this land has been ploughed and cereals grown. These are mainly barley and wheat. Other soils suitable for growing these crops are loams (mixtures of clay and sand) on top of clay.

Figure 2.13. *Winter wheat growing*
The wheat was planted in the autumn and will be ready for harvesting in the summer. Cereal production per area of land <u>doubled</u> in Britain from 1975 to 1985.

Climate
Arable crops grow best where the temperature is high in the summer. The south and east have average temperatures of about 17°C in July, whereas western areas are a degree or so cooler. There is less cloud in the south and east, so the days are more sunny. This helps to ripen the cereals. In the winter it is much colder there than in the western side of Britain (January temperature 4°–5°C in the east; 6°–8°C in the south-west). This means more frost. Frost helps to break up the soil and also discourages insects. Rainfall is about 700 mm in southern and eastern England. Much of East Anglia has over 600 mm of rain each year. There is plenty of moisture to encourage the growth of crops.

Farming · ARABLE FARMING

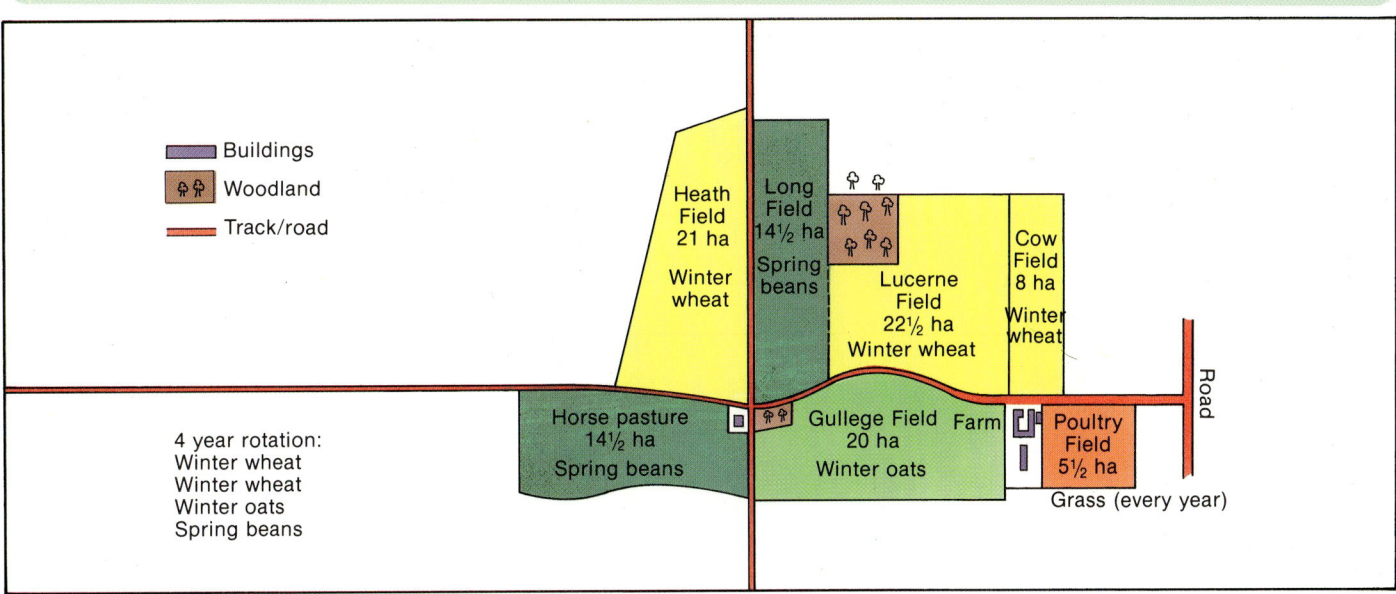

Figure 2.14. *Plan of Imberhorne Farm (see text on p. 44)*
The total size of the farm is 110 ha. What is the total size of all the fields? What is in the remaining area?

Figure 2.15. *Wheat being grown on the farm*
Note the absence of hedges. Why do you think the farmer has removed these?

Figure 2.16. *Oilseed rape*
Oilseed rape has a very distinctive colour and is a crop which has become increasingly popular in recent years.

IMBERHORNE FARM

An example of an arable farm is Imberhorne Farm, near the border of Sussex and Kent in south-east England (see Figure 2.14). The farm is just over 100 metres above sea level, on land gently sloping to the north-east. The soil is a loamy clay. It has been drained over the last six years by a series of underground pipes in the soil. There are seven fields, ranging in size from 5½ to 22½ ha. No animals are kept on the farm. The smallest field is Poultry Field, which is kept in grass each year. Sometimes the farmer lets a neighbour graze cattle and horses on it. Once it was a football pitch. Now it is used mainly for hay, which is cut twice a year and sold.

All the other fields have arable crops. These are grown in a crop rotation taking four years. Each field has wheat for two years, followed by oats, and finally spring beans in the fourth year. In the fifth year, the rotation starts again with wheat. Wheat and oats take out a lot of goodness while they are growing. The spring beans put some goodness back into the soil by way of valuable nutrients. The beans help to keep the soil rich.

The farmer has to add a carefully prepared and balanced fertiliser to keep the soil in its best condition.

The land is ploughed and planted with winter crops in the autumn. These crops are harvested mid-summer the following year. Spring beans are planted in March and April ready for harvesting in September. The beans and oats are used for animal fodder. Part of the wheat is also used for animal fodder. The rest is sent to flour mills and then sold to local bakeries.

It is a long time since there were any animals on this farm. As a result, the farmer has removed many of the hedges to make his fields bigger. For example, Heath Field was once three smaller fields. In 1982, two fields joined to become Lucerne Field. Removing hedges provides the farmer with more space to grow crops. It also cuts down the number of pests and diseases which can reduce the yield of the crops. But some people worry about the effect on wildlife when hedges are taken away. Sometimes trees and small copses are chopped down as well so that farmers can have bigger fields.

Imberhorne Farm is small in comparison with some farms in East Anglia. This is one of the most important areas of cereal growing in the country.

In East Anglia, arable farming is much more important than dairying. The main crops are wheat, barley and sugar beet. These three crops take up 95% of all the crop land. Sugar beet is a root crop grown to produce sugar as well as fodder for animals. There is a fixed price for sugar beet, so the farmer knows how much he is going to earn before he sells the crop. The barley and wheat prices can be expected to vary when the crops are sold at market.

East Anglia is generally flat and gently undulating. The land lends itself to mechanisation in ploughing, seeding and harvesting.

Figure 2.17. *Barley ready for harvesting*
Less barley is grown now than before 1980. Look at the wheat shown on p. 42 and compare the two crops.

Farming · ARABLE FARMING

Exercises

1. (a) What is meant by the term crop rotation?
   (b) Why does a farmer carry out crop rotation?
   (c) (i) Copy the following table.

	Heath Field	Long Field	Lucerne Field	Cow Field	Horse Field	Gullege Field	Poultry Field
Year 1							
2							
3							
4							

   (ii) Fill in the crops for year one as they are shown on the map of the farm in Figure 2.14.
   (iii) Assuming the farmer follows the four-year rotation, complete the rest of the table by working out what you think will be growing in each field.

2. (a) Describe the shape of the fields. How does this help the farmer?
   (b) Suggest why the farmer has removed most of the hedges. Who might be against arable farmers removing hedges and why?

3. The farmer wants to make as much money as possible from the crops. Use the following information to work out how much money is received for the crops listed.

   Yield: Wheat – 3 tonnes per hectare
   Barley – 3 tonnes per hectare
   Beans – 2 tonnes per hectare
   Price on market:
   Wheat – £120 per tonne
   Barley – £100 per tonne
   Beans – £80 per tonne

   To work out the figure for each crop:
   (i) Find the size of the field in hectares.
   (ii) Work out the yield of the field by multiplying the yield per hectare by the size of the field in hectares.
   (iii) Work out the money the farmer will gain by multiplying the total yield of each crop by its price per tonne.

4. What will be the farmer's other costs (such as transport to market, fertilisers, pesticides etc.)?

Figure 2.18. *Sugar beet growing*

Questions

1. What are the main conditions that favour arable farming in East Anglia? Consider: (a) relief; (b) soils; (c) climate; and (d) mechanisation.

2. (a) What is the difference between cash crops and fodder crops?
   (b) Find out how the following crops can be used:
   wheat   barley
   sugar beet
   oil seed rape

3. In Norfolk, the land is divided as follows:

Cereals	210 000 ha
Potatoes	52 000 ha
Fruit and vegetables	28 000 ha
Grass	82 000 ha
Rough grassland	24 000 ha
Total	396 000 ha

   (a) Draw either a bar graph or a pie chart (divided circle) to show the way in which the land is used in Norfolk.
   (b) Describe the way in which the farmland is used in Norfolk, using either the table, or the graph you have drawn.

Hill Farming

About one third of Britain's farmland is in upland or mountain areas. Much of the land is over 300 m above sea level, with steep-sided valleys. Temperatures are lower than in the lowlands. The rainfall is generally much heavier, and there is a lot more frost and snow in winter. With poor grassland at lower levels and rough moorland higher up, few animals can survive in these conditions. Acidic and peaty soils are another problem and make any type of farming in hilly areas very difficult.

Farmers in some upland areas, like the Pennines, manage to make a living from rearing sheep. They sell wool and breed lambs. The lambs can be sold for meat after they have been fattened on lowland farms.

But most hill farmers still find it very difficult to make a reasonable living. They can only increase profits by keeping more animals. However, these have to be fed in winter, which means improving grassland and growing fodder crops. Some farmers in upland areas of eastern Britain, where there is a lower rainfall, have some success in growing hardier strains of barley for winter feeding. These farmers obtain a subsidy to burn the vegetation off the moorland, plough it, and add fertiliser to the peaty soil. But even if the land can be made more productive, life on hill farms is still very hard. Most farms are very isolated. Some have only recently been connected to mains electricity supplies. Living conditions are often poor. During the winter months, farmers and their families can be cut off by snow for weeks on end. Nowadays, fewer people, especially young people, want to live such a hard, lonely life. Hill farms in some upland areas now lie empty and deserted.

Figure 2.20 *The moorland here was burnt and seeded with grass, providing grazing land.*

○ BROWNSETT FARM

Brownsett Farm is a hill sheep farm. It is on the edge of the Peak District National Park, on the border between Staffordshire and Derbyshire. Brownsett is typical of many upland farms in this area.

The farm is just under 100 ha in size, and has 500 sheep. The land varies in height between 230 m and 400 m above sea level. The relief is of a west facing slope, steep in places, with a stream flowing in the bottom of the valley. Much of the land above 360 m is moorland. No attempt has been made to turn it into grassland as the soil is very thin and there are several rock outcrops.

Apart from the moorland, the remainder of the farm is given over to grassland. Some new grass is sown each year. Hay and root crops (turnips) were once grown on the lowest slopes of the valley. Now the land is too wet and stoney to plough. Little grows on the moorland, although there is some

Figure 2.19. *The remoteness of Brownsett Farm*
Beyond the farm is moorland on the hills. What do you think are the main problems of living there?

bracken and heather on the flatter areas. Above 360 m, the rocks are sandstone. Below this height there is shale. Shale makes heavy soil, and is therefore difficult to plough.

Sheep are reared for wool and lambs. The farmer shears his own sheep. The animals are out all the year round. As they have to be fed during the winter, they are brought closer to the farm during these months. When it is cold and snowy, they huddle behind stone walls for protection. The lambs are sent to market and sold to lowland farmers for fattening. The farmer also keeps two cows for milk.

The farmer earns his living from the sale of wool and lambs. He also gets a subsidy from the Government. This is based on the number of sheep that he keeps. Although the farm is isolated, his nearest market is only 10 km away. But this can seem a long way, particularly during the winter when the farm is often cut off by huge snowdrifts.

During the worst of the winter, when even the landrover cannot move, the farmer uses ponies to pull a sledge around the farm. This helps him to keep an eye on his sheep. Breeding ponies is one way that the farmer manages to make some extra money. He also does some welding, and helps to repair machinery and tractors for other farmers. In return, his neighbours help him when it is sheep shearing time.

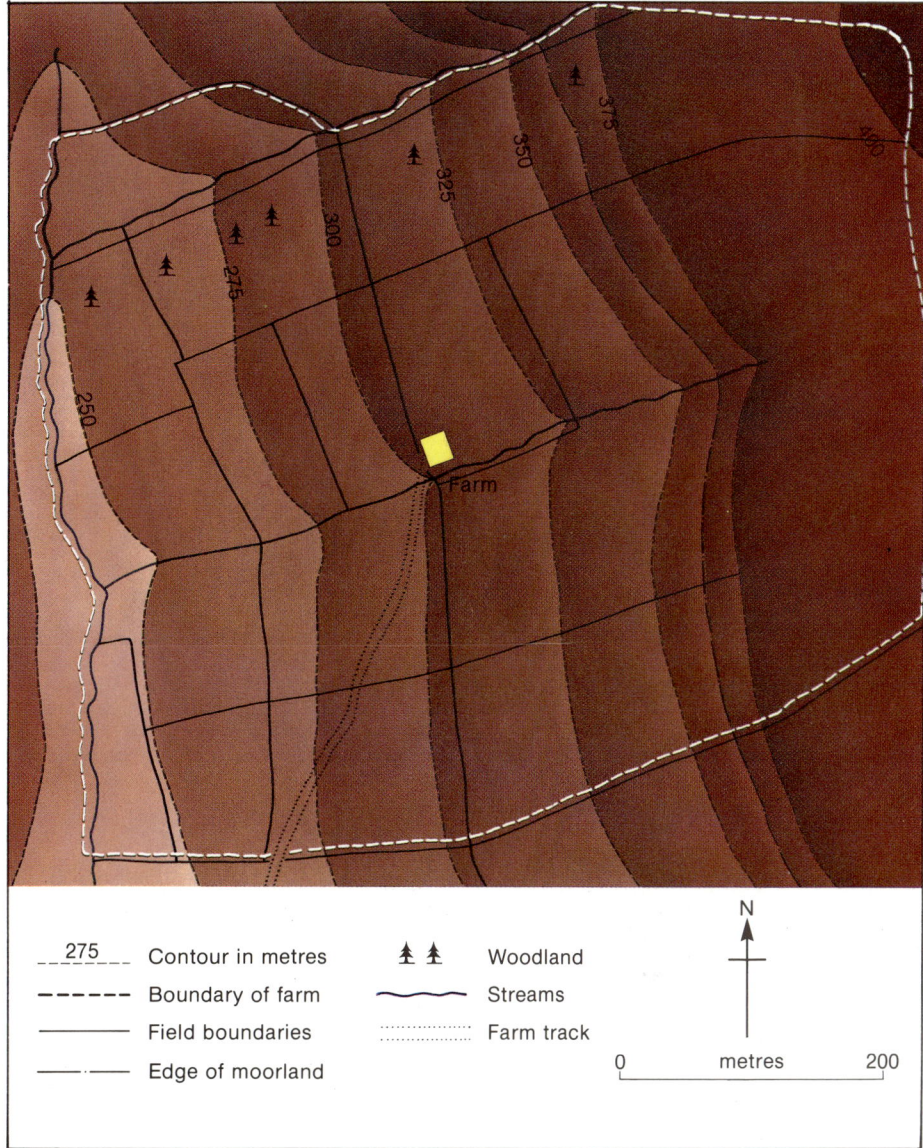

Figure 2.22. *Plan of Brownsett Farm*
(a) *Suggest why the land in the east is moorland.*
(b) *Describe the location of the woodland.*
(c) *Some fields near the river were used for the growing of crops for animal feed. Why do you think these fields were most suitable? Suggest reasons why the farmer has now stopped growing these crops.*

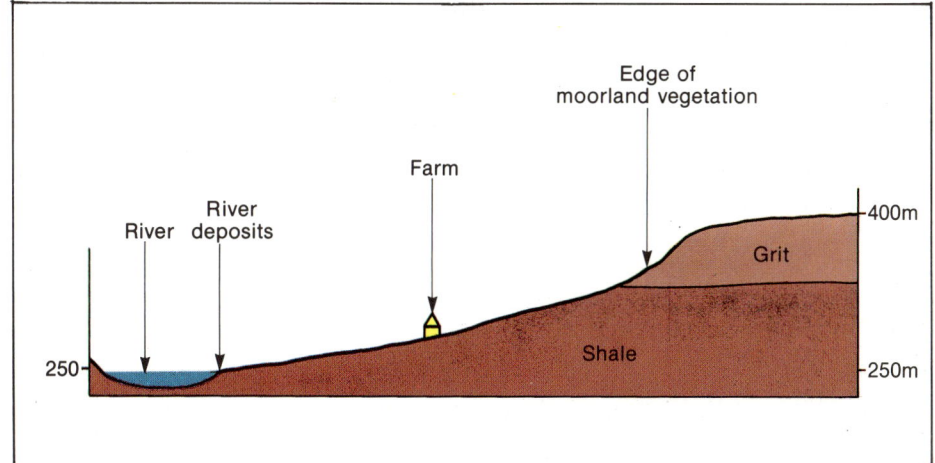

Figure 2.21. *Section through the farm from west to east*

Exercises

1 Life in the upland hill farms is very difficult. Describe the difficulties a hill farmer faces in terms of:
(a) the weather;
(b) the land;
(c) earning enough money.

2 In view of these difficulties, describe and suggest reasons why there are still a large number of hill sheep farmers.

Market Gardening

The climate along the south coast of England is enjoyed by farmers as well as holidaymakers. Using glasshouses, farmers can grow crops like tomatoes, lettuces and flowers all the year round. Market gardening has become a very important type of farming in this part of the country. This type of farming takes place in a nursery.

One example of a market garden lies 4 km north of the holiday resort of Bognor Regis. It is at the foot of the line of hills known as the South Downs.

The nursery covers 0.6 ha, on a flat site. All the crops are grown in glasshouses or under plastic. The area under glass is 0.4 ha, and 0.2 ha is covered by plastic sheeting, which is cheaper and also easier to replace than glass. The nurseryman actually owns a lot more land, but it is too expensive to develop it all as a market garden.

There are three main growing areas. Two are under glass and one is under plastic. All are heated by steam. The nursery produces lettuces, cucumbers, sweet peas, chrysanthemums and astamarias. The last three crops listed are flowers. Until 1976, a large number of carnations were grown each year. After that it became impossible for the nursery to compete with cheaper imports.

The soil is a clay loam, but contains many flints. Each year weeds and diseases have to be removed. The soil is sterilised using either gas, which is cheaper, or steam. Afterwards, a sample is sent to a nearby research station. The nurseryman can then find out what minerals and nutrients he should add to the soil to get it into perfect condition for growing new crops.

All the crops need a constant temperature of between 15°C and 16°C. They also need water. This is run through each greenhouse in pipes, like the steam. The amount of water, or 'rainfall', is very carefully controlled to ensure maximum growth.

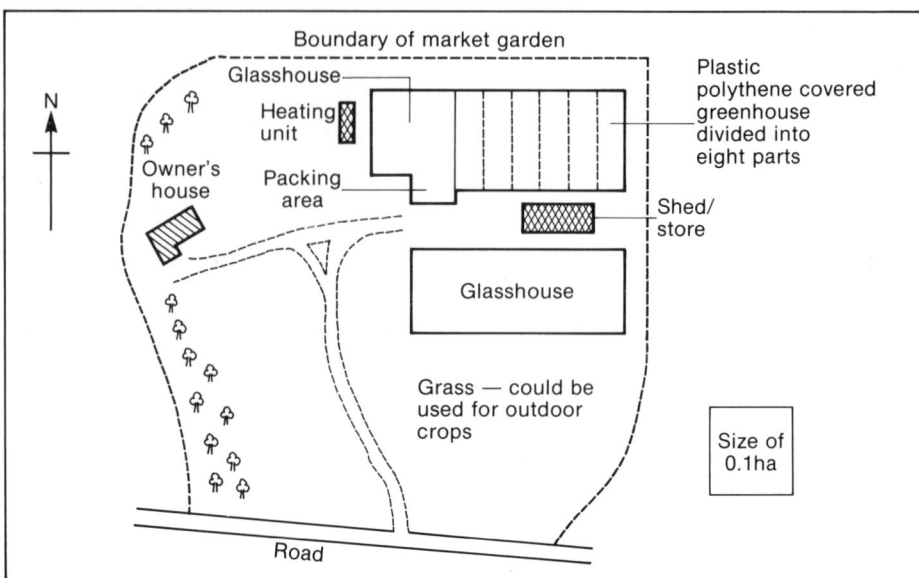

Figure 2.23. *Plan of the market garden*
The produce is grown from seed. When ready it can be sold locally or throughout the country. What crops could be grown in the area of grass?

The nurseryman sends his crops to market. Most of his customers are in the south of England. Some of his more specialised produce, like chrysanthemums, is sent throughout the country – as far north as Aberdeen. Local buyers collect from him to supply greengrocers and markets in the area. The grower also uses a marketing organisation to send crops to other parts of the country. This is based in Southampton. A lorry collects produce from the nursery on a regular basis.

Until recently, the high cost of transport was the nurseryman's main worry. Now he is more concerned about the high cost of oil. He needs large amounts of oil to heat his greenhouses. Growers in other countries, like the Netherlands, have received large Government subsidies. Nurserymen in Britain are given only 5p per gallon of oil towards the cost of their fuel. However, the subsidies granted in some Common Market countries are now being phased out. British growers should soon be able to compete better with European imports. In the past these imports have often been cheaper than home grown produce because of the fuel subsidies.

The grower also has to pay for insecticides, and labour. He sprays his produce regularly to protect it from pests and diseases. Apart from the nurseryman and his son, two full-time and four part-time workers are employed to tend and pick the crops.

The only machinery used on the nursery is a small tractor. This has been specially designed to work in greenhouses. The tractor reduces labour costs because the soil no longer has to be dug by hand.

During recent years, competition from abroad has forced many changes in this type of farming. The subsidies given to foreign growers, like those in the Netherlands, have been the main reason. For example, market gardeners in Britain can no longer compete and make a profit growing crops like carnations. Our nurseryman is always experimenting with new crops. This is why he has introduced astamarias.

Farming · MARKET GARDENING

Figure 2.24. *Cauliflowers being picked in Lincolnshire*

Figure 2.26. *A field of lettuces*

Figure 2.25. *The glasshouse*
Glasshouses vary in size as this photograph shows. What are the advantages of growing crops in this glasshouse?

Questions

1 Write your own account of this farm, using the following headings:
(a) The artificial climate.
(b) The care of the soil.
(c) The crops grown.
(d) The main costs involved (soil, glasshouses, heating, labour, transport).

2 What are the advantages of growing salad crops and flowers throughout the year?

Map reading questions

1 Look at Map 1 on page 123.
(a) Find the farm at grid reference 444733. Suggest reasons why dairying is likely to be the main type of farming.
(b) Describe the site of Holme Farm in grid square 4577. What are the main problems of this farm?
(c) Much of the area on this map was once farmland. Describe the other types of land use that have taken up valuable farmland. Approximately how much land has been taken up with the oil refinery?

2 Look at Map 2 on page 124. In grid square 9109 there are several farms. Describe the type of farming you would expect to find here. Give reasons for your answer. What are the main problems that these farmers face?

3 Look at Map 3 on page 125.
(a) Suggest reasons why there is little farming in grid square 0213.
(b) Briefly describe the site of Horseylane Farm (062138) and Manor Farm (080170). Suggest what type of farming may take place at Manor Farm, giving your reasons.

SECTION 3 · *Industry*

Mention the word industry, and most people will think of factories, steelworks and shipyards, producing anything from nuts and bolts to giant oil tankers. It is this manufacturing, or secondary, industry which we shall be looking at in this section. There are other kinds of industry as well. Some extract raw materials, like coal mining or supply food, like agriculture. These are called primary industries. Others which provide services, like tourism, are known as tertiary industries.

We buy many of the goods made in British factories ourselves. But we also have to sell our products to other countries as exports. Although there are some raw materials in Britain, like coal and oil, we still have to import others, like rubber, from abroad. We also have to import much of our food. The money we receive from our exports helps us to pay for these imports.

Iron and Steel

Iron is one of the most common elements in the earth's crust. Its uses are enormous. Without iron, the world would seem a very different place. As long ago as the fifteenth century, iron was being made in Sussex, the Forest of Dean and North Yorkshire by primitive methods (Figure 3.1). In the eighteenth century, the Industrial Revolution increased the uses of iron and made it into one of the world's most important metals.

Iron is obtained by heating iron ore (mainly iron oxide) in a blast furnace. This produces pig iron. Afterwards, it can be heated again to produce steel. Some of the steel is then rolled into sheets in a rolling mill.

○ IRON ORE

One of the raw materials needed to make iron and steel is iron ore. Iron ore is found in many different parts of the world. In some areas, such as northern Sweden and West Africa, there are very rich deposits (i.e. over 70% of the rock is iron). In other places, the quality of the ore is much poorer (i.e. less than 50% of the rock is iron). All of the iron ore found in Britain has an iron content of less than 40%. The rest is waste. As it costs as much to transport ore containing a lot of waste as it does to carry ore with a

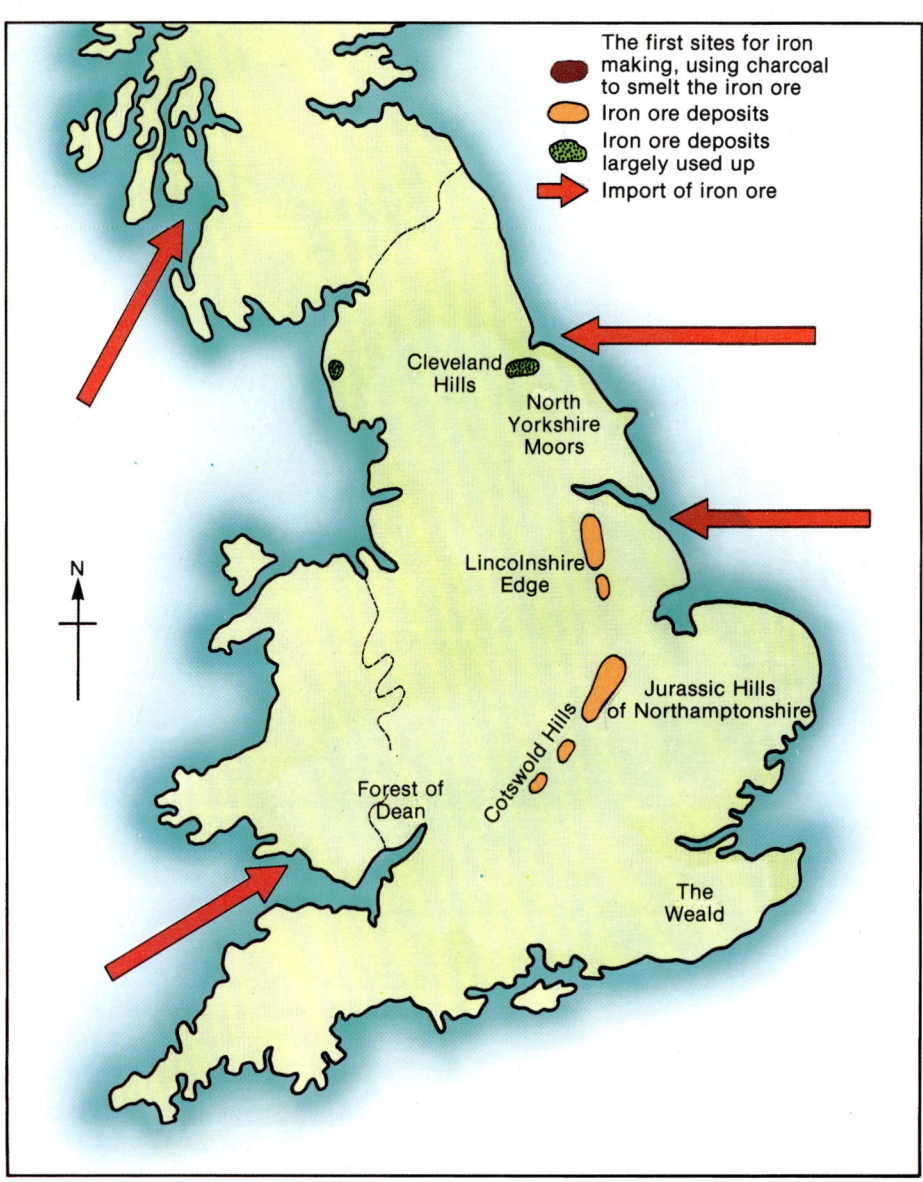

Figure 3.1. *Location of main iron ore deposits in Britain*

Industry · IRON AND STEEL

high iron content, many iron and steel makers have looked to other countries for their supplies. It is often easier, and much cheaper, to import better quality iron ore from countries like Sweden, West Africa and Brazil. Ore imports from these countries have been increasing steadily since 1950, as Figure 3.2 shows.

Britain still produces iron ore. The main area is in Lincolnshire. Open cast quarrying methods are used to take the ore out of the ground, after the overlying rocks (called the overburden) have been removed. Large mechanical shovels collect the iron ore, and load it on to waiting dumper trucks or trains. The overburden is stored elsewhere, so that when all the ore has been extracted, the rocks and soil can be replaced, and the area landscaped. Land is then soon fit for farming.

○ THE RAW MATERIALS OF A BLAST FURNACE

Figure 3.5 shows the three raw materials needed to produce pig iron. Coke is obtained by heating coal in special coke ovens. Good coking coal is still found in Britain but much must be imported.

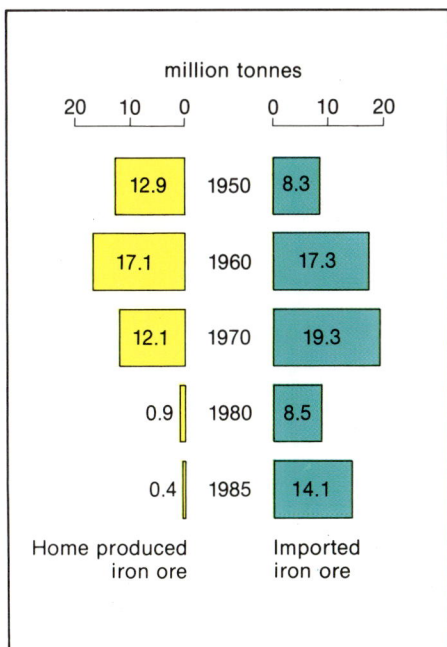

Figure 3.2. *Amount of iron ore from Britain and overseas (million tonnes)*

Figure 3.3 *Quarrying for iron ore*
Describe how the iron ore is being quarried. Include information on the size of the quarry and the equipment being used.

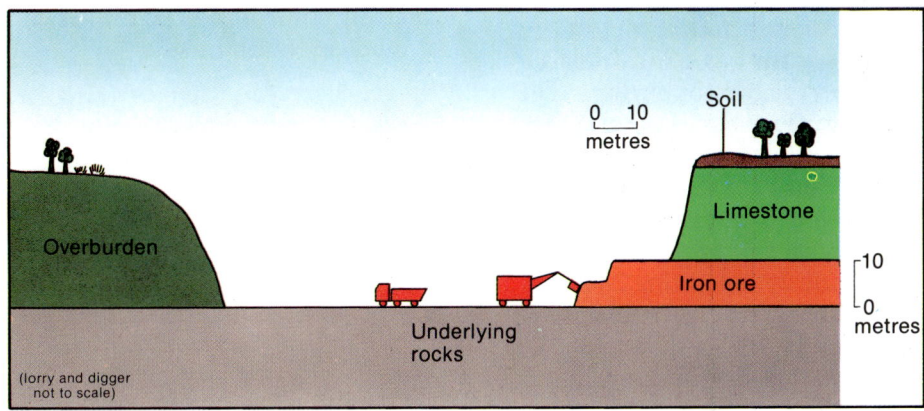

Figure 3.4. *Section through an open cast quarry*
The limestone and the soil together form the 'overburden' which is moved elsewhere when quarrying for iron ore.
The overburden is taken to a used part of the quarry and used for 'backfilling'. Later the land can be used again for agricultural purposes.

Questions

1 With reference to the diagram (Figure 3.4):
 (a) State how thick the layer of iron ore is.
 (b) State how thick the layer of limestone is.
 (c) What rock forms part of the overburden?
 (d) What happens to the overburden when quarrying is taking place?
 (e) When quarrying has finished, what happens to the overburden then?

2 How do you think a community might be affected by quarrying for iron ore? Consider what the advantages and/or disadvantages might be for the following:
 (a) a nearby iron smelter;
 (b) the local council;
 (c) a farmer;
 (d) a nearby housing estate.

3 Describe carefully how an area of land is quarried.

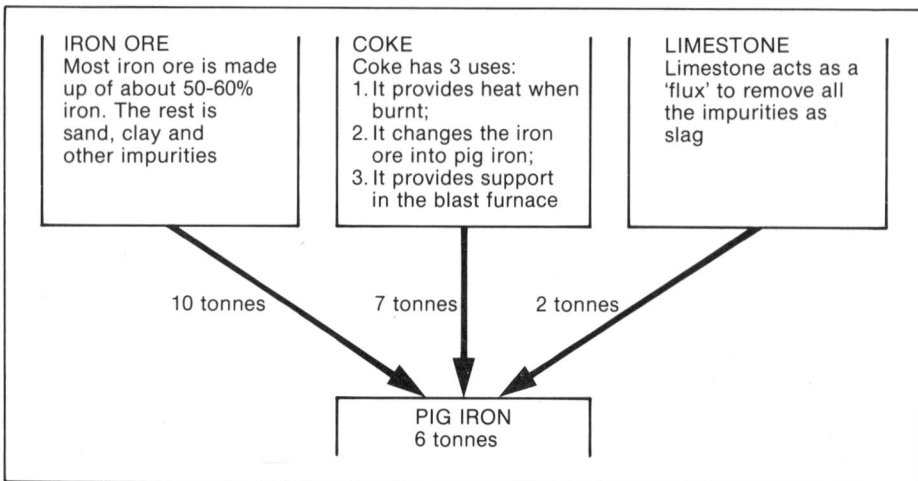

Figure 3.5. *Raw materials*

Look at Figure 3.8. It can turn a mixture of molten iron and scrap into steel in 40 minutes. The molten iron and scrap is loaded into a furnace (known as a converter), and a water cooled oxygen lance is lowered into the mixture. Pure oxygen is blown through the lance at high speed. The steel is produced as the oxygen combines with the unwanted chemicals, such as carbon. As this method is very fast and efficient, it is gradually replacing the older ways of making steel, like the open hearth.

○ THE BLAST FURNACE

Figure 3.6 shows a diagram of a blast furnace. The raw materials are fed into the top of the furnace. Most blast furnaces are about 70 metres high and 11 metres or more at the base. An air blast increases the heat of burning. The temperature inside rises to over 1800°C. Molten iron is collected near the base, at regular intervals. It is then sent to be made into steel.

The blast furnace is making iron continuously, operating 24 hours a day, seven days a week. Eventually, the refractory bricks lining the inside start to deteriorate and the furnace has to stop. This happens about once every several years. The refractory bricks are replaced and the process starts up again.

Molten iron is 90–95% pure. The remainder is made up of impurities.

○ MAKING STEEL

There are three main ways of making steel:

The open hearth furnace

Molten iron, scrap and limestone are fed (or charged) into shallow hearths. Flames are swept over the mixture. The open hearth furnace is a slow process, taking up to ten hours to change 350 tonnes of iron into steel.

The electric arc furnace

Electrodes pass a powerful electric current into the scrap steel. This heats it and refines it. The electric arc is suitable for high grade steels because the temperature can be carefully controlled. 150 tonnes can be made in four hours.

The basic oxygen furnace

This is the most modern process.

○ INTEGRATED WORKS

Much of the steel produced goes to a rolling mill. After it has been rolled into sheets and other shapes, it is sold to customers who will use it for products such as car bodies.

Before it can be rolled, the steel has to be heated yet again. Reheating would take a great deal of energy if the metal was allowed to cool down too much as it was moved from one stage to the next during the hot rolling processes.

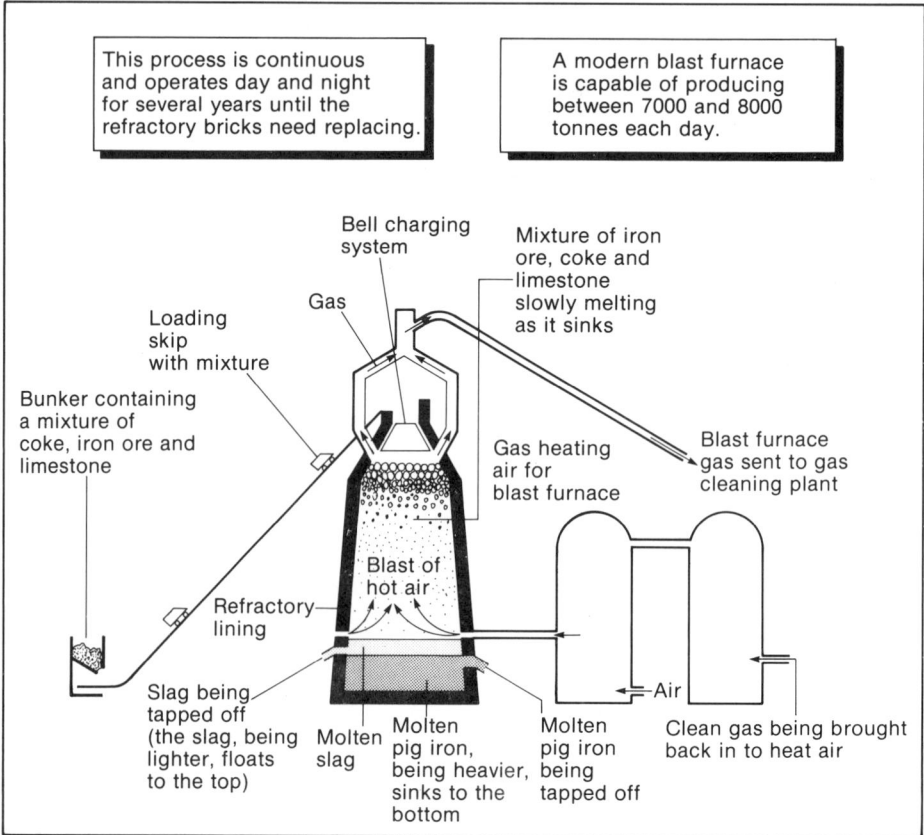

Figure 3.6. *A blast furnace*

Industry · IRON AND STEEL

Figure 3.7. *Basic oxygen steel plant, Scunthorpe. Hot metal is being poured into one of the three 300-tonne furnaces.*

Most iron and steel works are therefore integrated. The blast furnace, steel furnace and rolling mill are all on one site. Iron can be moved to the steelworks, and the steel to the rolling mill, very quickly. Reheating costs are kept low. Transport costs are also reduced. The whole operation is much more efficient than it would be if each part of the works were on a different site.

○ IRON AND STEEL IN CLEVELAND

Iron making began in Cleveland over 100 years ago. By 1850, the industry had grown quite large, using iron ore from the Cleveland Hills and coal from the Durham Coalfield. But it was soon realised that the local ore was of poor quality. It also contained a great deal of phosphorus. This was a particular disadvantage when steel making was introduced. It was very difficult to make good quality steel out of ore containing phosphorus. As local supplies were beginning to dwindle anyway, the iron and steel makers decided to import ore from abroad. Iron ore from countries such as Sweden, Spain and North Africa has a high iron content and no phosphorus.

Up until the middle of this century, most of the steelworks in Cleveland were quite small. Much of the equipment they used was old. The industry found it difficult to introduce new ways of making iron and steel. In 1972, it was decided that several small works should close, to be replaced by large new steel making plants, like Lackenby.

Lackenby began making steel in the mid-1970s. At the same time, work started on what was to be Britain's and Europe's largest blast furnace at Redcar. Such modern works use the latest technology. Some processes are so new that they have not yet been used elsewhere.

Here, iron and steel making can be divided into two parts:
(a) the making of iron at Redcar;
(b) the making and rolling of steel at Lackenby.

At Redcar, iron ore is imported from overseas in bulk carriers. The ore is unloaded at the ore terminal, then moved by conveyor belt to stock piles and stored until required.

The blast furnace at Redcar makes 10 000 tonnes of iron per day, and works around the clock every day of the year. The furnace produces large amounts of gas during the production process. One third of this is recycled through the furnace, and the rest goes to supply a power station built on the site. This is a good example of energy conservation.

Iron produced at Redcar is taken to the steelworks at Lackenby every day.

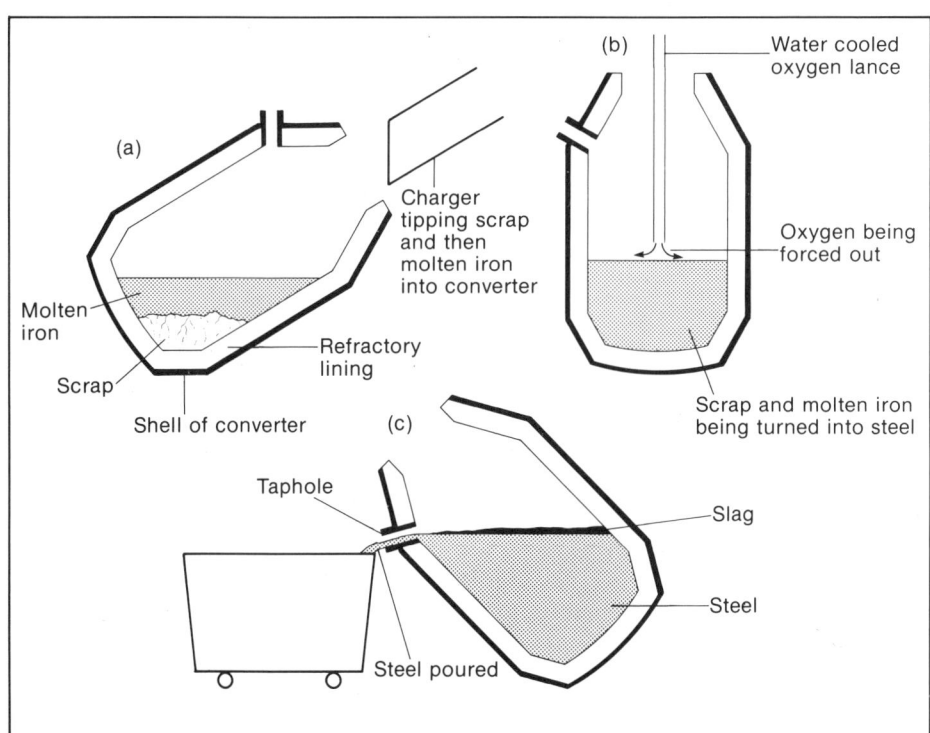

Figure 3.8. *The basic oxygen furnace*
(a) *Converter is tipped to allow scrap and molten iron to be fed in.*
(b) *Converter is returned to its upright position and the oxygen lance is plunged in, creating what steelworks call a 'blow'.*
(c) *Converter is tipped the other way to allow the molten steel to be poured out. The slag is finally tipped out.*

The British Isles: Themes and Case Studies

Figure 3.9. *Steelmaking*

Some of the finished steel is exported from Cleveland. A large amount is also transported by road and rail to customers in other parts of Britain.

The steel making complex at Redcar/Lackenby is a coastal site. Supplies of iron ore can be imported easily. Finished goods can also be readily exported, through Teesport.

○ CHANGES IN IRON AND STEEL

During the last 30 years, many changes have taken place in the iron and steel industry. Demand for British steel has declined, so some steelworks have had to close. Others have been modernised, and can now produce iron and steel more efficiently, with fewer workers. Many steelworkers have lost their jobs. Whole communities, have suffered from cuts and closures.

A fall in demand for steel from British shipbuilders is one reason for the decline in the industry. Fewer ships are being built. Many of those that are ordered are completed in foreign shipyards.

Also, less steel is being used in heavy engineering. This has added to the problems of iron and steel making in Britain.

Today, bulk steel production is centred on five large integrated works. These are Llanwern and Port Talbot in South Wales, Scunthorpe and Redcar in England, and Ravenscraig in Scotland. In addition, there are smaller works in the Sheffield area making more specialised steels such as alloy and stainless.

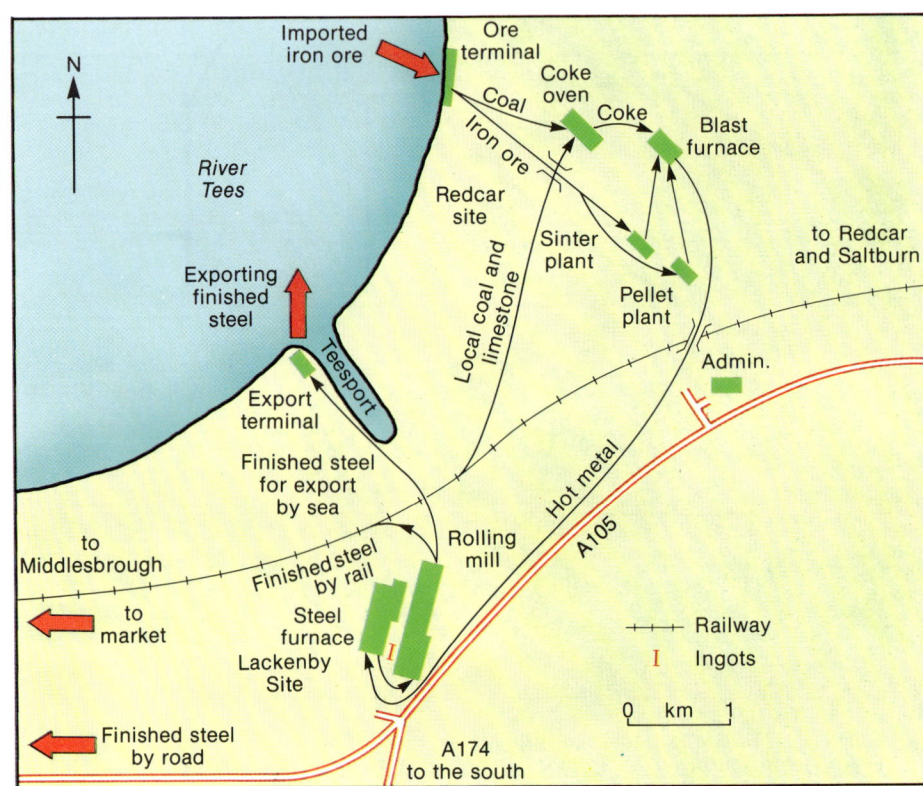

Figure 3.10. *Layout of iron and steel making at Redcar and Lackenby Try to write your own account of what is happening in this diagram.*

Industry · IRON AND STEEL

Figure 3.11. *Large steelworks in operation in the 1940s*
There were also many other smaller works.

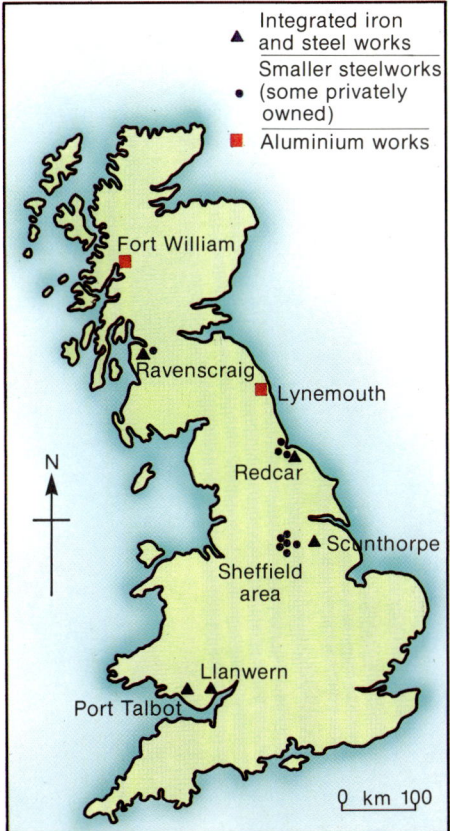

Figure 3.12. *Iron and steel works, 1980*

○ ALUMINIUM

Aluminium is a very light and strong metal. It is used in a great many products, ranging from saucepans to aircraft parts. Its properties make it almost as useful as iron.

Aluminium is a common element. It occurs in many rocks near to the earth's surface or just below. But it is only present in concentrated deposits in a few areas. These deposits are called bauxite (an oxide of aluminium).

Changing bauxite into aluminium requires a great deal of electricity. Over 10 000 kW of electricity is needed to make 1 tonne of aluminium. (Your kettle at home might use 2 kW each time it boils.)

As so much electricity is needed, aluminium smelters are built close to sources of power. In Britain, for instance, the Fort William smelter in Scotland is near a hydroelectric power station. The plant at Lynemouth in north-east England is close to a coal mine. They are shown on Figure 3.12.

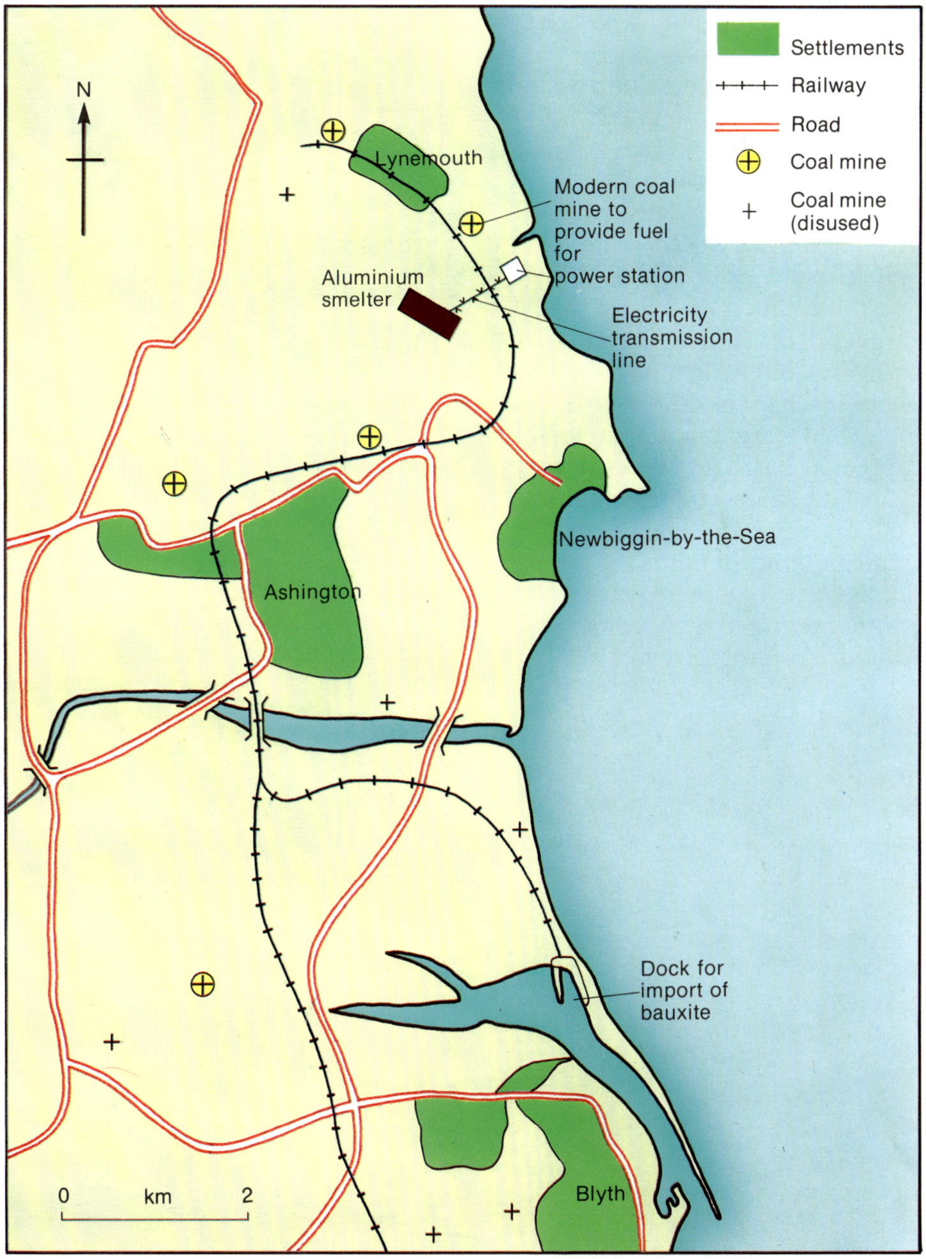

Figure 3.13. *Lynemouth aluminium smelter*
Write an account of why an aluminium smelter has been built here. (Government grants played an important role.)

Oil Refining

○ STANLOW

Stanlow is an oil refinery on the south side of the Mersey Estuary. Situated alongside the Manchester Ship Canal, the refinery is 13 km north east of Chester and 20 km from Liverpool.

The refinery opened in 1922 as a small plant. It gradually increased in size until 1974, when a new crude distillation unit was built. This increased output by 85%. Stanlow can now process 18 million tonnes of crude oil a year. It is one of the largest refineries in Britain today.

The site at Stanlow was originally chosen because of its position next to the Manchester Ship Canal. Ships could travel right up to the refinery to unload their cargoes of oil. This kept transport costs low. But soon tankers got too big to reach Stanlow. By 1949, many had started to unload their cargoes at Eastham Docks. Twenty years later, a new oil terminal was built at Tranmere. Today, only small amounts of crude oil are unloaded at Tranmere because another terminal was opened at Amlwich on Angelsey in 1976.

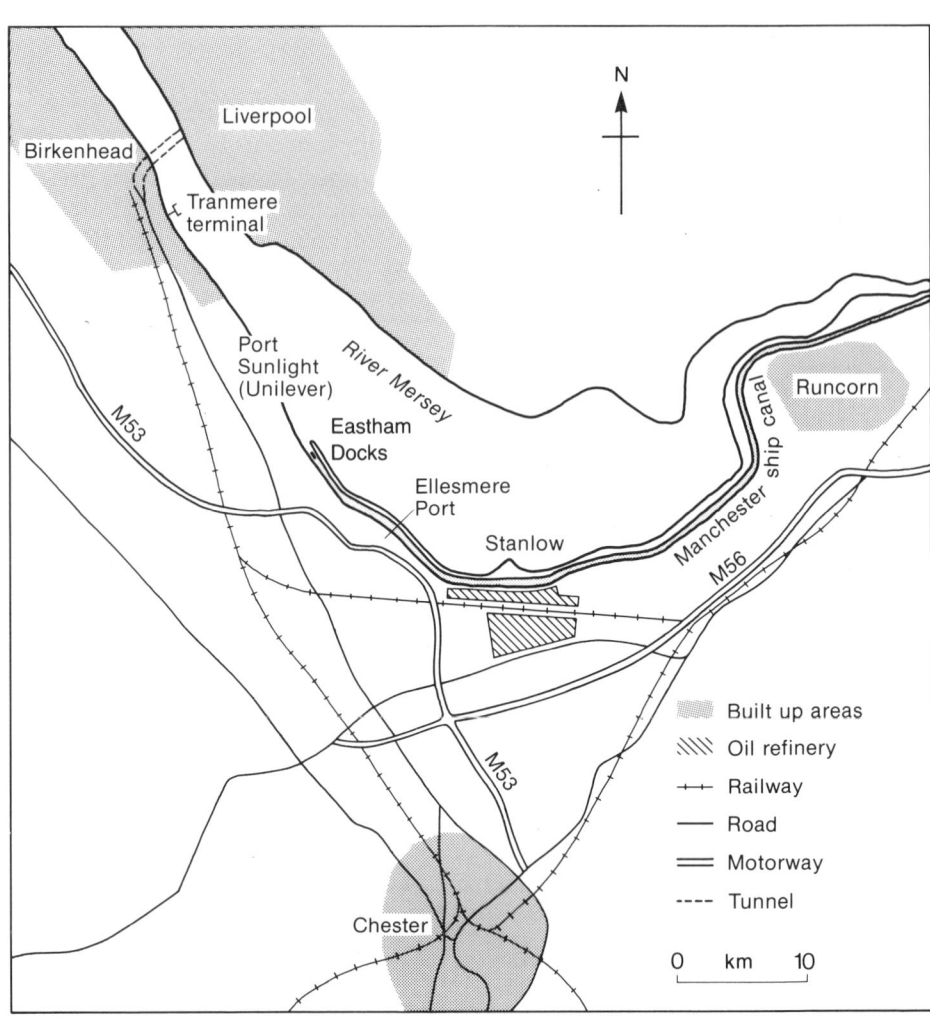

Figure 3.15. *The position of Stanlow oil refinery*

Figure 3.14. *Stanlow oil refinery*

Industry · OIL REFINING

Amlwich was designed to handle ULCCs (Ultra Large Crude Carriers). These enormous ships can carry up to 500 000 tonnes of oil. Crude oil unloaded at Amlwich is then taken across North Wales to Stanlow, by pipeline. This is a distance of over 132 km. The use of the pipeline has greatly reduced congestion amongst ships in the River Mersey.

Each day Stanlow produces enough petrol for 100 000 cars to travel from London to Glasgow and back; over 11 million litres of aircraft fuel, sufficient for Concorde to fly to America and back 32 times; enough oil for 17 power stations; sufficient bitumen to surface 2 km of motorway; enough wax to make four million candles; also industrial oils, oil for central heating, and oil for ships.

All of the oil refined at Stanlow has to be transported to customers. Fifty per cent of its production goes by pipeline to the power stations at Ince and Billingham; to Unilever, for detergents; and to the chemical industry at Carrington. Some goes by pipeline to a terminal midway between Liverpool and Manchester. Other oil is transported by another pipeline to the Midlands.

Twenty per cent is sent to smaller British ports by coastal tankers from Eastham Docks. The rest goes by road and rail.

Any oil refinery requires huge amounts of water for cooling purposes. Stanlow draws most of what it needs from the Manchester Ship Canal. After water has been used, it is treated and purified and returned to the Canal.

Four thousand people are employed at Stanlow. This figure includes office staff as well as production workers.

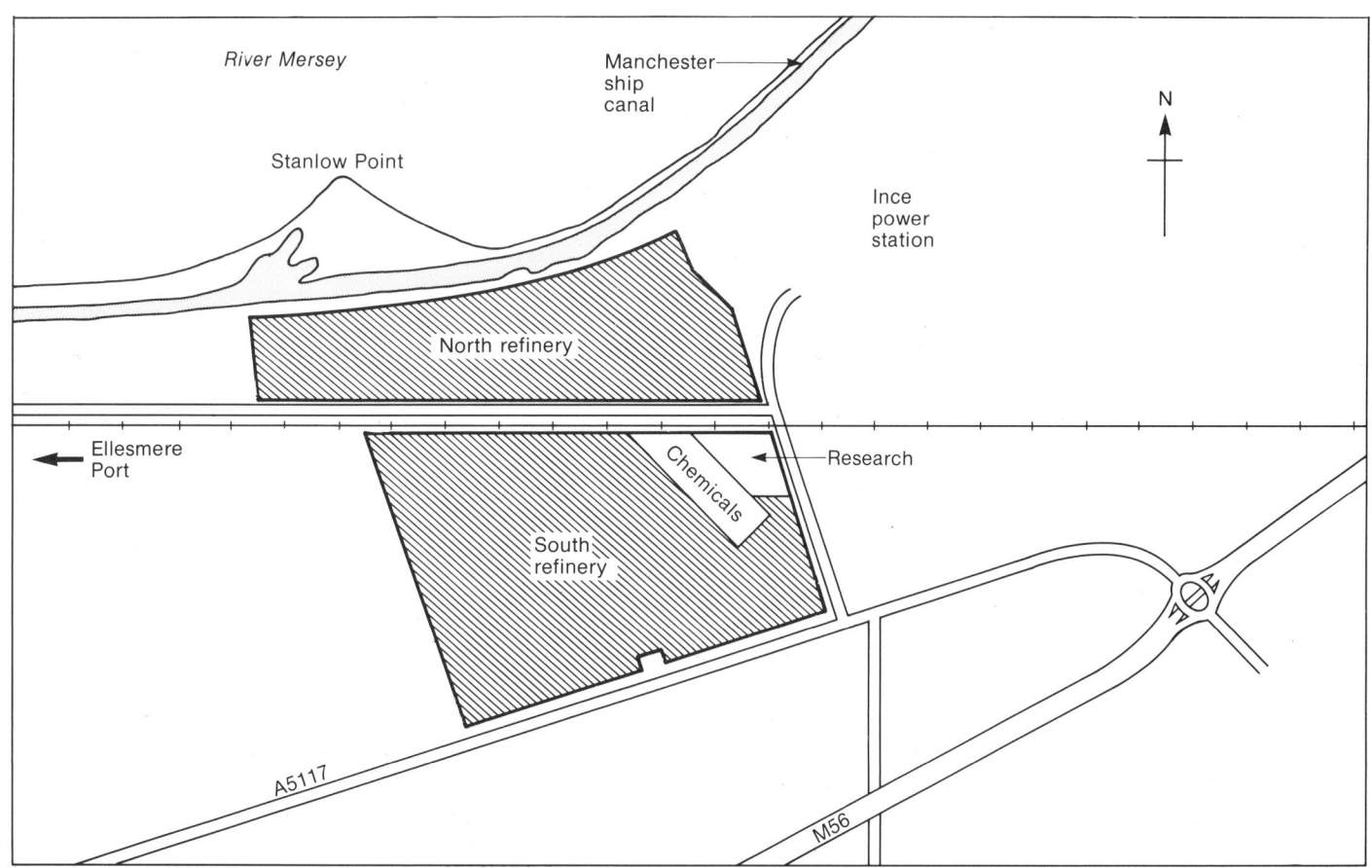

Figure 3.16. *The refineries at Stanlow*

Questions

1 List the important products produced at oil refineries like Stanlow.

2 Explain why much of the crude oil refined at Stanlow is obtained from North Wales by pipeline.

3 Draw a pie diagram to show the methods used to transport the finished products from the Stanlow refinery. Explain why pipelines are very important for taking the finished products to markets.

4 Study the maps in Figures 3.15 and 3.16 and describe why Stanlow is a suitable location for an oil refinery.

Chemicals

The chemical industry uses large quantities of a great variety of raw materials. The most important include salt, gypsum and limestone. Products made from oil are called petrochemicals. We shall look at these later.

Most of the raw materials needed to make chemicals are available in Britain. There are deposits of coal, salt, gypsum and limestone in various parts of the country. Salt, for instance, is used in the production of acids and alkalis, PVC, bleaching powder and disinfectants. Other materials, like sulphur for making sulphuric acid, and nitrates for fertilisers, have to be imported.

Chemical works are usually located:
(a) close to a source of necessary raw materials;
(b) close to good supplies of water;
(c) close to a port, especially if raw materials like sulphur phosphates and nitrates have to be imported;
(d) near to an efficient transport system, so that raw materials can be transported to the works both by road and rail. Many raw materials are very bulky.

Figure 3.17 shows the area of Cheshire and south Lancashire. From the map, you should be able to list the factors that have made the area important as a centre for chemicals. In addition, labour is available from towns like Liverpool. South Lancashire also provides a good market for chemical products, like paint.

Salt is extracted from the rocks underground in Cheshire. Today, Sandbach and Middlewich are the main areas from which salt is obtained. Some of the salt is transported to the main chemical centres by brine pipeline.

Merseyside is the major centre of Unilever's detergent and speciality chemicals businesses in the UK. The chemical companies manufacture fatty acids, glycerines, silicates and flavours for many different uses.

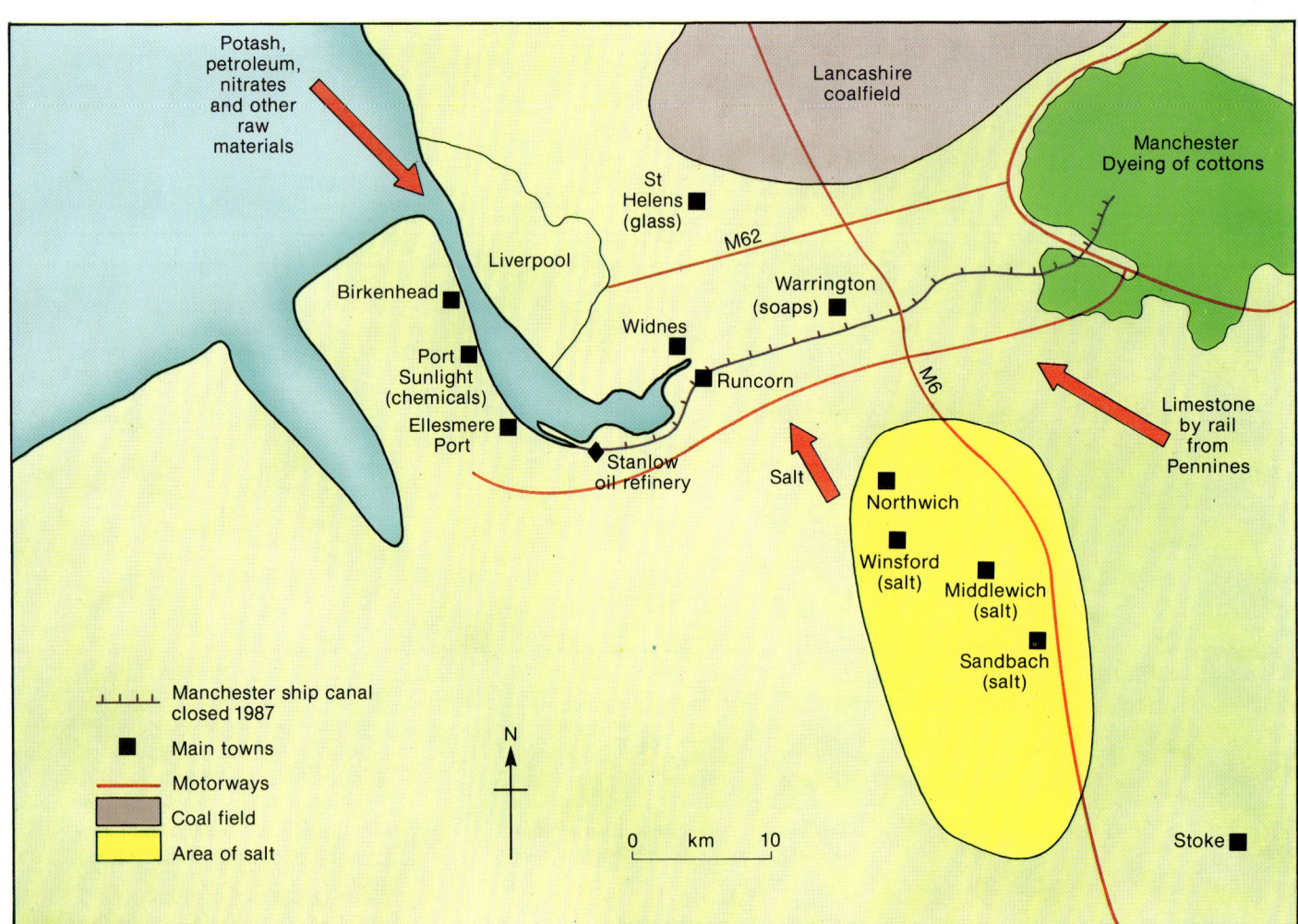

Figure 3.17. *The chemical industry in Cheshire.*
Salt is transported by pipeline and by lorry from the saltfield to the towns in the north. Runcorn has a chemical works belonging to ICI.

Industry · CHEMICALS

Figure 3.18. *The inside of PPF International's flavours plant at Bromborough, Merseyside*

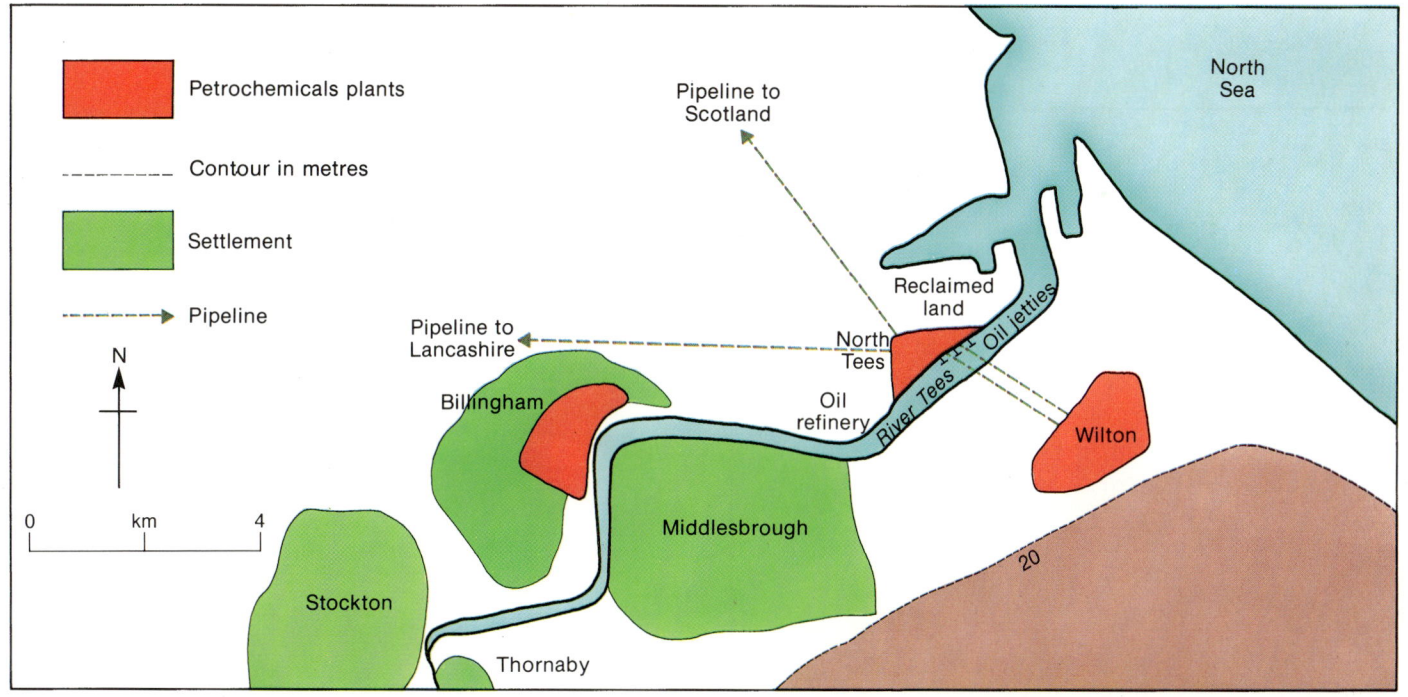

Figure 3.19. *Cleveland*
Using the map, suggest two likely reasons why the North Tees site was chosen for petrochemical plants. What other factors were important (see map on p. 9 and information about raw materials on p. 61).

Questions

1 Study Figure 3.17. List the raw materials used in the chemical industry in Cheshire and describe their source. What forms of transport are used?

2 Roads are used for transporting many of the finished products in the chemical industry. Suggest reasons why road transport is so important.

3 Write an account of where chemical industries are usually located giving specific examples where possible.

○ PETROCHEMICALS

As the name suggests, petrochemicals are chemicals produced from petroleum. The industry developed in the 1950s as part of the chemical industry. Before then, petroleum or oil was virtually all made into petrol for cars. The rest was wasted. Petrochemicals became important when it was discovered that this waste could be turned into a wide range of products.

Today, this range of products includes nylon, terylene, plastic, polythene, PVC and perspex. Petrochemicals are even used to make some animal foodstuffs.

The petrochemical industry also makes new varieties of plastics for special needs. Examples are certain types of 'perspex' made from polypropylene (which can withstand very high impact) and plastics that will not melt even up to a temperature of 200°C. Some plastics are being used widely in space technology for their special properties.

The petrochemical industry relies heavily on petroleum. This has to be imported, from abroad or from the North Sea. The industry also uses a great deal of water. Most large petrochemical works are located close to or are on the coast. The sites chosen are usually very large.

The industry makes a wide variety of products and needs plenty of space. Most

Figure 3.20. The ICI petrochemicals plant and offices at Wilton, Cleveland. The River Tees can be seen at the top of the photo. Describe the layout and type of building in the foreground. How have planners tried to make the area more pleasant? From about where on the map (Figure 3.19) did the photographer take this picture?

Industry · CHEMICALS

petrochemical plants are built on cheap land which would be unsuitable for housing or agriculture. However, the sites must not be too isolated because the industry needs a large workforce. One complex can employ as many as 10 000 people.

One of the largest areas in Britain for petrochemicals is Cleveland. The industry lies on either side of the River Tees, near Middlesbrough.

Billingham, Wilton and North Tees are the three most important petrochemical sites in Cleveland.

Billingham was the first chemical plant opened in Cleveland. Production began in the 1920s. All the essential raw materials were available locally, i.e. coal, water and salt. The site was also chosen for its good road and railway links.

Production began at Wilton in 1949. The plant has grown much larger since that time to keep up with the increasing demand for petrochemicals. There is still plenty of room for further expansion. Only two thirds of the 800 ha site has so far been developed.

The site at North Tees is an oil refinery. It has the advantage of deep water jetties. This means petroleum can be unloaded in the shelter of the Tees Estuary. Much of the petroleum used comes from the North Sea.

Wilton and Billingham are joined by two pipelines running under the River Tees. These carry raw materials and products between the two sites. Two further pipelines provide links with Grangemouth in Scotland and with the chemical works in Cheshire across the Pennines.

All these sites are sheltered by the Tees Estuary. The River Tees also supplies all the water that the industry requires. Most of the workforce the industry needs comes from the surrounding towns of Middlesbrough, Stockton and Billingham. The number of people employed has declined in recent years.

Figure 3.21. Wilton and Billingham began many years ago. When factories such as this are planned today, many people can be affected in different ways. Above are some imaginary people giving their points of view. Who seems to be in favour of the proposed new factory? Who seems to be opposed? Can you think of any other people or groups who might have an opinion? Try holding your own planning meeting in class. Have six people pretend to be the six characters shown. Each should stand and speak for 2 minutes giving their view for or against the plan.

Questions

1 Explain what the term 'petrochemicals' means. Name six products from the petrochemical industry. For each product, give an example of its everyday use.

2 Explain why plastics are such an important product. What advantage do they have over other materials, such as steel?

3 List the main raw materials used in the petrochemical industry.

4 Describe the factors that determine the site of a new petrochemical industry. Who might object to its site and why?

The Car Industry

At the turn of the century, there were several UK firms making cars. Some examples were Riley, Rover and Sunbeam – who began as companies making bicycles. They manufactured small numbers of cars, making both engine and body in their own workshops. Other components (car parts) were supplied by different firms. Many of these companies were located in the West Midlands. Gradually the area around Birmingham became very important for the car industry.

Later, William Morris started to assemble cars at Cowley, near Oxford. He bought the bodies and engines from other firms, instead of making these parts in his factory. During the 1930s, there was a big increase in the number of cars manufactured. Car makers tried to build models that ordinary people could afford. For example, the Austin Seven sold for between £100–£225 in the 1930s.

Figure 3.23. *The very first Austin Seven, which was sold to the public in 1923. Production of various models continued until 1939. Nearly 300 000 were made altogether.*

Figure 3.22. *Workers at Morris Motors at Cowley hammer on the upholstery and body trimming, 1939.*

Figure 3.24. *Major car producing areas in Britain*

Industry · CAR INDUSTRY

In America, Henry Ford had improved the moving assembly line. Large numbers of the same model of car could be produced as workers stayed in one place and cars moved along the line. Each worker did the same job over and over again. Car manufacturers in Britain tended to change the design of their cars more often than the Americans.

There was some delay before the assembly line was introduced in Britain. Gradually it became popular as more and more people wanted to buy cars. The assembly lines needed a large area of factory space. Car factories grew much larger and were built on the edges of towns and cities, where they had more room to expand.

During the 1950s there were many different car makers, all competing with one another. Then they started to merge. In 1951, Austin and Morris joined together to form the British Motor Corporation (BMC). This was the beginning of Austin Rover.

Today there are four large firms producing cars in Britain. These are:
(a) Austin Rover, based at Birmingham and Oxford;
(b) Ford at Dagenham and Halewood;
(c) Vauxhall (General Motors), at Luton and Ellesmere Port;
(d) Talbot at Coventry.

Ford and General Motors are American firms. They have many factories in Europe which manufacture parts that are then brought to Britain for assembly. Some British factories also make components which are assembled abroad. Both American companies produce some models entirely abroad, and import complete cars into Britain.

Figure 3.25. *The Japanese Nissan Motor Co. Ltd opened its first car plant in the UK in 1986. The Sunderland, Tyne and Wear plant is located on what was a disused airport. Under Phase One, the British workforce has been assembling cars from body panels and parts shipped mainly from Japan. The plan is for more and more actual production in the UK.*

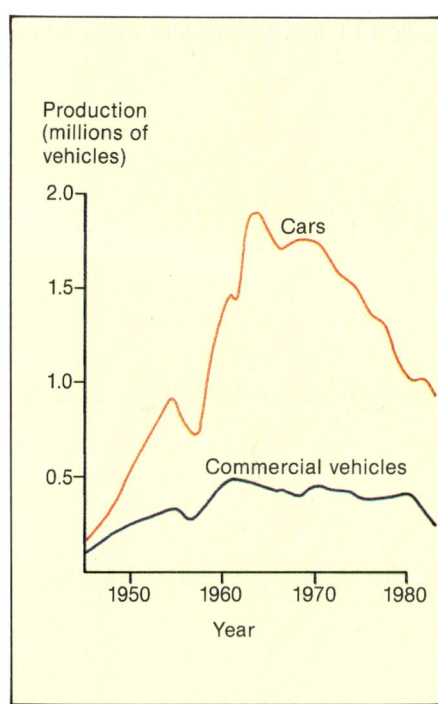

Figure 3.26. *Motor vehicle production in Britain since 1945*
What has been the trend of (a) cars and (b) commercial vehicles since the mid 1960s? Suggest reasons for these trends.

Map reading question

On Map 1, page 123, locate the motor car works in the north west of the map. What factors have been important in siting a car industry here? Why do you think grid square 4677 would be unsuitable for this type of industry?

LONGBRIDGE

Longbridge is one of British Leyland's largest car producing factories. It lies 14 km south-west of Birmingham. It is situated on the boundary between the West Midlands, and the counties of Herefordshire and Worcester.

The various parts of the factory cover a large area of land. The site is 2 km from north to south. Being fairly flat, it has been easy to construct the many buildings needed by a modern car plant.

Longbridge has other advantages as a site for car production. It is beside the A38 main road, which runs from the north-east to the south-west of the country. The plant is only 5 km from the M5 motorway. This provides a fast road link with South Wales and the south-west of England. In the opposite direction the M5 connects with the M6 to Lancashire and Scotland. A new motorway planned to skirt the south of Birmingham will also benefit Longbridge. It will provide better transport links with other Austin Rover plants at Solihull, and Cowley, near Oxford. Alongside the factory is the main railway line from the south-west and South Wales to the north-east.

Every car produced consists of many different components (or parts). A large proportion of those used at Longbridge are manufactured in the West Midlands. Figure 3.30 shows some of the different components that are needed and where they are made. The fast transport links around Longbridge are of considerable importance in obtaining components made in other parts of the country.

Car production needs a large labour force. 18 000 people work at Longbridge. The nearby West Midlands provides a workforce with many different skills for the car industry.

Figure 3.27. *Aerial view of Longbridge* The main building in the front of the photograph is where car bodies for the Metro are made. The main assembly area is to the top right of the photograph.

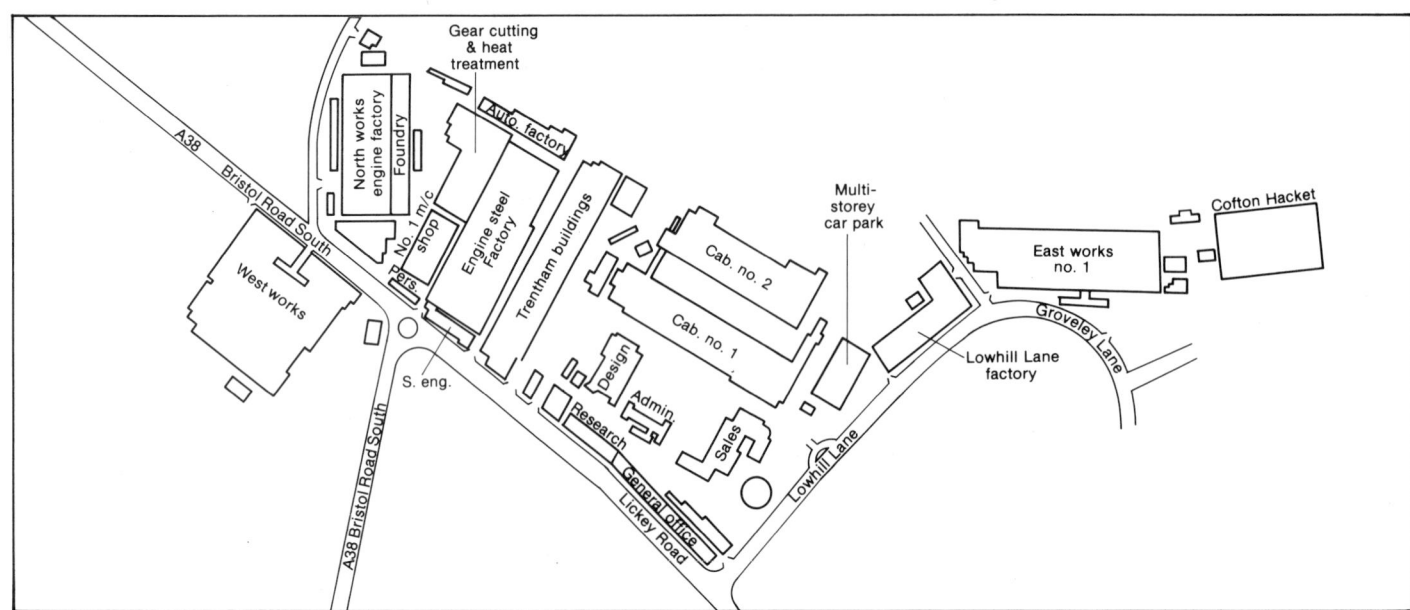

Figure 3.28. *Longbridge*

Industry · CAR INDUSTRY

Figure 3.29. *The assembly line*
Robots weld the body of the car. The assembly line is totally automated. Very few workers are needed, but in the past this process involved larger numbers of workers. What are the advantages of producing cars using automation?

Figure 3.30. *Some of the parts used to make a car*
The diagram shows the sources of parts for an Austin Rover model. On this car, 98% of the parts are British made.

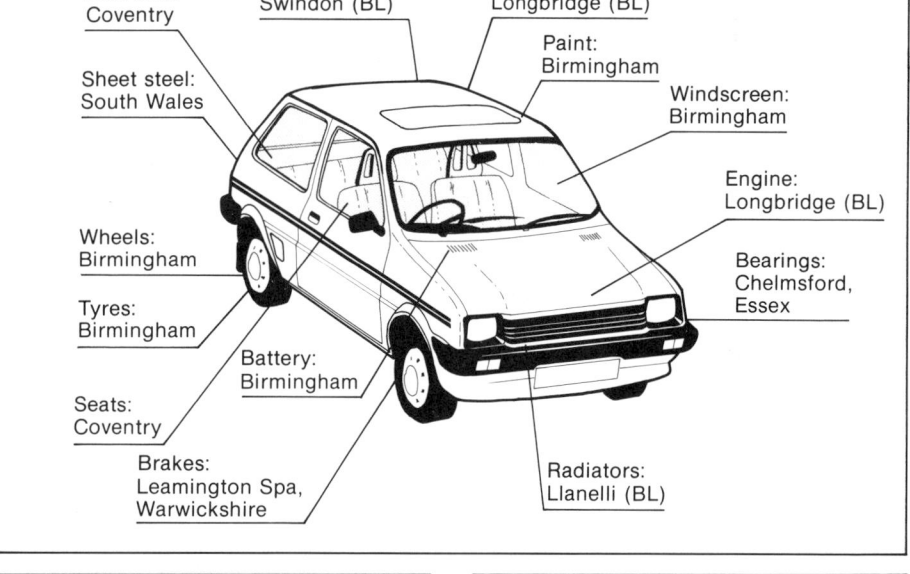

Soft trim: Coventry
Body pressings: Swindon (BL)
Assembly: Longbridge (BL)
Paint: Birmingham
Sheet steel: South Wales
Windscreen: Birmingham
Engine: Longbridge (BL)
Wheels: Birmingham
Bearings: Chelmsford, Essex
Tyres: Birmingham
Battery: Birmingham
Seats: Coventry
Radiators: Llanelli (BL)
Brakes: Leamington Spa, Warwickshire

Figure 3.31. *The final inspection*
The cars are carefully checked and cleaned before being sent to garages.

Questions

1 Name the four main car producers in Britain and state where they are based.

2 Why is it an advantage to build large numbers of cars on an assembly line?

3 Imagine you want to set up your own car factory. What factors do you have to consider before choosing your site?

4 List the advantages that Longbridge has for producing cars.

Textiles

The textile industry includes the making of clothing, soft furnishings, bedding and other materials with fibres. The most important 'natural' fibres include wool, cotton and linen. But many synthetic fibres are now being used as well, such as nylon, rayon and terylene.

Britain was once the leading producer of textiles. But the industry has declined rapidly in recent years. Many other countries now make textiles, often more cheaply and efficiently than in Britain.

In Britain, people made cloth in their own homes many centuries ago. It was mainly for their own use, but some was bought and sold in local markets. Textile making began to develop more rapidly during the sixteenth century, when weavers from Belgium began to settle in East Anglia and south-east England. Norwich, in East Anglia, became the centre of a small yet important textile industry.

In the middle of the eighteenth century, woollens were being made on the eastern side of the Pennines – around Leeds and Huddersfield. Woollen makers were attracted to sites that had:
(a) plenty of local wool available, from sheep grazing on the moorlands;
(b) fast flowing rivers and streams, so that water wheels could be used to power the early forms of machinery;
(c) soft water, found in areas lying on millstone grit, which could be used for 'softening' the wool.

The valley of the River Aire, upstream from Leeds, was an important location for early woollen mills. When steam driven machinery was introduced, wool making gradually moved downstream, closer to the coalfield.

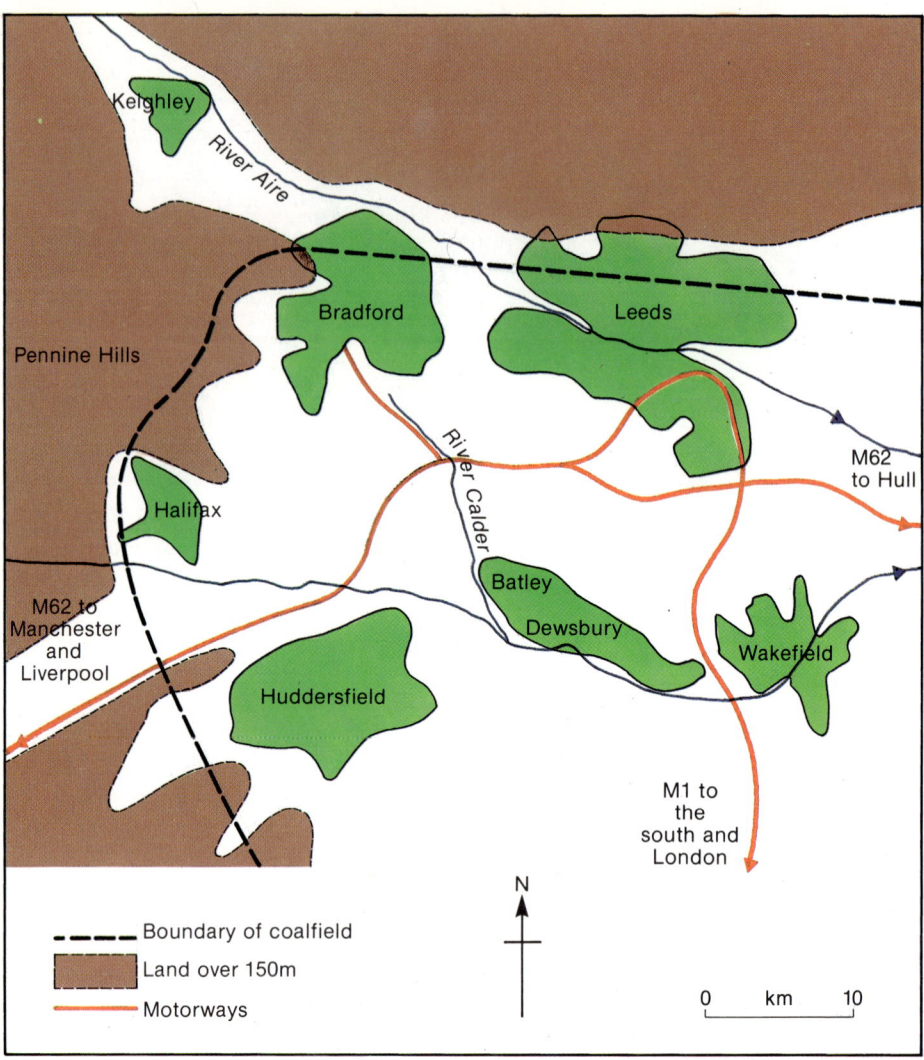

Figure 3.32. *The Yorkshire woollen area*

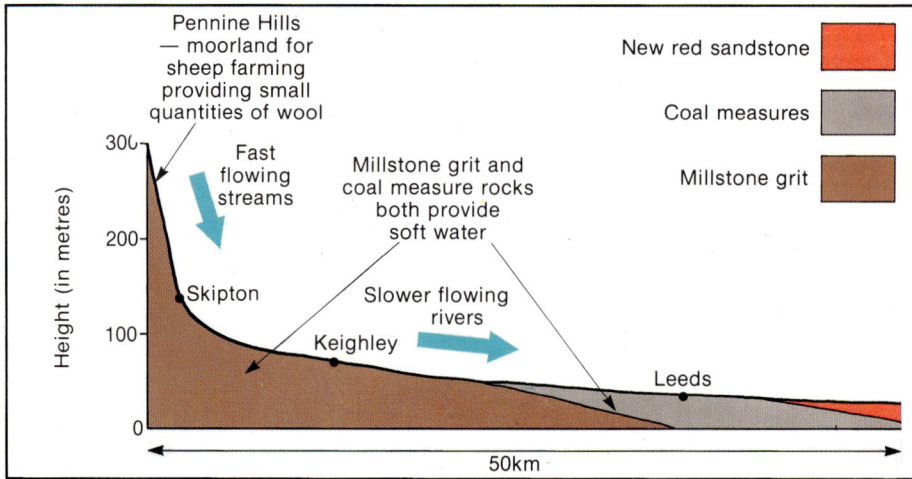

Figure 3.33. *Long profile of the river Aire from the Pennines to beyond Leeds*
From the section, state
(a) why the woollen industry started around Skipton and Keighley;
(b) why it moved towards Leeds;
(c) why it is unlikely to move further downstream beyond Leeds.

Industry · TEXTILES

Leeds eventually became the centre for woollens. The city is 80 km from the port of Hull. This was fortunate when the quantity of wool from the Pennine sheep declined. The woollen industry might have collapsed if it had not been possible to import alternative supplies through Hull docks. The imports were also cheaper than the home product. Most of the wool was imported from Australia. Much is still imported today through Hull, and some through London as well.

Wool passes through several processes before it becomes woollen cloth. It has to be washed and combed and then it is spun into yarn. After the yarn is woven into cloth, the cloth has to be dyed and 'finished' before leaving the mill.

Huddersfield and Bradford are two of the most important cloth making towns. Leeds is the main administrative centre for the woollen industry. It is also famous for tailoring. Dewsbury is known for its 'shoddy' (cloth made from wool and rags). Halifax tends to concentrate on making carpets. Together, these are the 'traditional' woollen towns.

Figure 3.35. *Spinning the wool*
The long lengths of wool (yarn) are twisted and wound onto bobbins.

Figure 3.36. *Weaving the wool*
Bobbins with yarns of different colours are used to form different patterned woollens.

Figure 3.34. *Wool scouring*
The wool has to be cleaned by washing. Scouring removes the dirt, sweat and grease. The clean wool is then dried in the chamber at the far end of the room.

Although there is competition from overseas, the woollen industry appears to be remaining fairly steady. However, it is possible that other synthetic fibres will replace wool to some extent. These synthetic fibres have similar properties to wool and can be produced more cheaply.

The woollen industry in Britain needs to become more efficient. Many firms have already **started** to integrate production by **introducing** spinning and weaving in the **same** mill, instead of on separate sites.

○ COTTON

South Lancashire is the main cotton producing area in Britain. It started, like Yorkshire, as a wool producing district but changed for a variety of reasons. Merchants depended on the port of Liverpool to import raw cotton and export cotton goods. During the last century, the cotton industry became very important. There were several reasons for this:

(a) the introduction of new machinery meant that large quantities of cotton goods could be produced in a short time;
(b) Liverpool was nearby. It was convenient for the import of raw cotton from America and the export of finished cotton goods;
(c) the South Lancashire coalfield provided a cheap source of power for machines.

Such was the rise of cotton that in 1913 about three quarters of all finished cotton was sold abroad. Cotton exports made up nearly a quarter of all Britain's earnings from goods sold to other countries.

Cotton passes through several processes before it becomes cotton cloth. After cleaning, the raw cotton looks like a mass of cotton wool. It then has to be spun into a fine yarn. Some towns have specialised in spinning and are known as the spinning towns. Bolton and Bury are two of the most important of these. Weaving is the next important process. The weaving towns include Blackburn and Burnley. These towns are in the north of the area. The final stage in the cotton manufacturing involves bleaching, dyeing and printing the material. Manchester is the main centre for this finishing process.

Since 1913, the cotton industry has steadily declined. There are three important reasons for this:

(a) a lack of investment in modern machinery. Many factories today are using very old machinery;
(b) competition from other parts of the world, particularly Hong Kong and India. Labour is cheaper in these countries and some of the machinery is more modern;
(c) a large increase in the amount of goods made from synthetic fibres.

Many cotton mills have been forced to close. Others have merged with each other to try and stay in business. Towns do not specialise in spinning or weaving as much as they used to. Many firms have also started to manufacture special cottons and even synthetic fibres to avoid closure.

In areas where high unemployment has been caused by cotton mills closing, efforts have been made to create new jobs. A range of new light industries has been encouraged to move into the old factories. Attempts have been made to re-train workers for employment in industries such as electrical engineering, or in local craft workshops.

○ SYNTHETIC FIBRES

During the last twenty years there have been many changes in textile

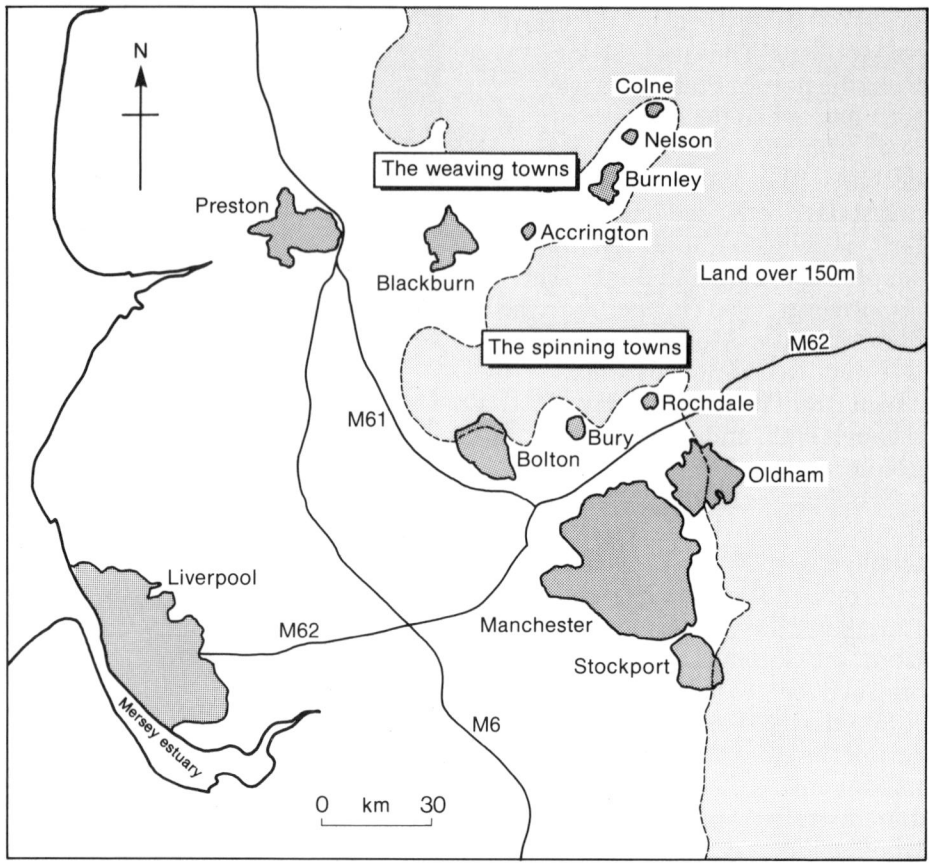

Figure 3.38. *The cotton towns*

Figure 3.37. *Cotton growing*

Industry · TEXTILES

manufacturing. The production of cotton in particular has gone down, but there has been an increase in the making of synthetic fibres. The synthetic fibres include nylon (made from oil) and rayon (made from wood cellulose). In addition, other varieties of synthetic fibres are produced by different firms. Many of these varieties have the same properties as each other.

Synthetic fibres have several advantages over natural fibres:
(a) they are usually stronger;
(b) they can be mixed with natural fibres, e.g. nylon with wool, terylene with worsted;
(c) some synthetic fibres have special properties, e.g. crease resistance or less tendency to

Originally, synthetics were produced in the traditional textile areas of Britain (Lancashire and Yorkshire). The declining production of wool and cotton coincided with the rise of synthetic fibres. In recent years, the synthetic textiles industry has become scattered throughout Britain.

This is because:
(a) the raw materials for the synthetic fibres are mainly based on oil, which is imported through a number of ports;
(b) the Government provided financial help to encourage new factories to be based in areas of high unemployment, e.g. the north-east of England and central Scotland;
(c) labour was available throughout the country;
(d) energy costs have become much the same throughout the country.

Consequently, synthetic fibres are now produced in Northern Ireland, Merseyside and South Wales, as well as in the older traditional areas of textile manufacturing.

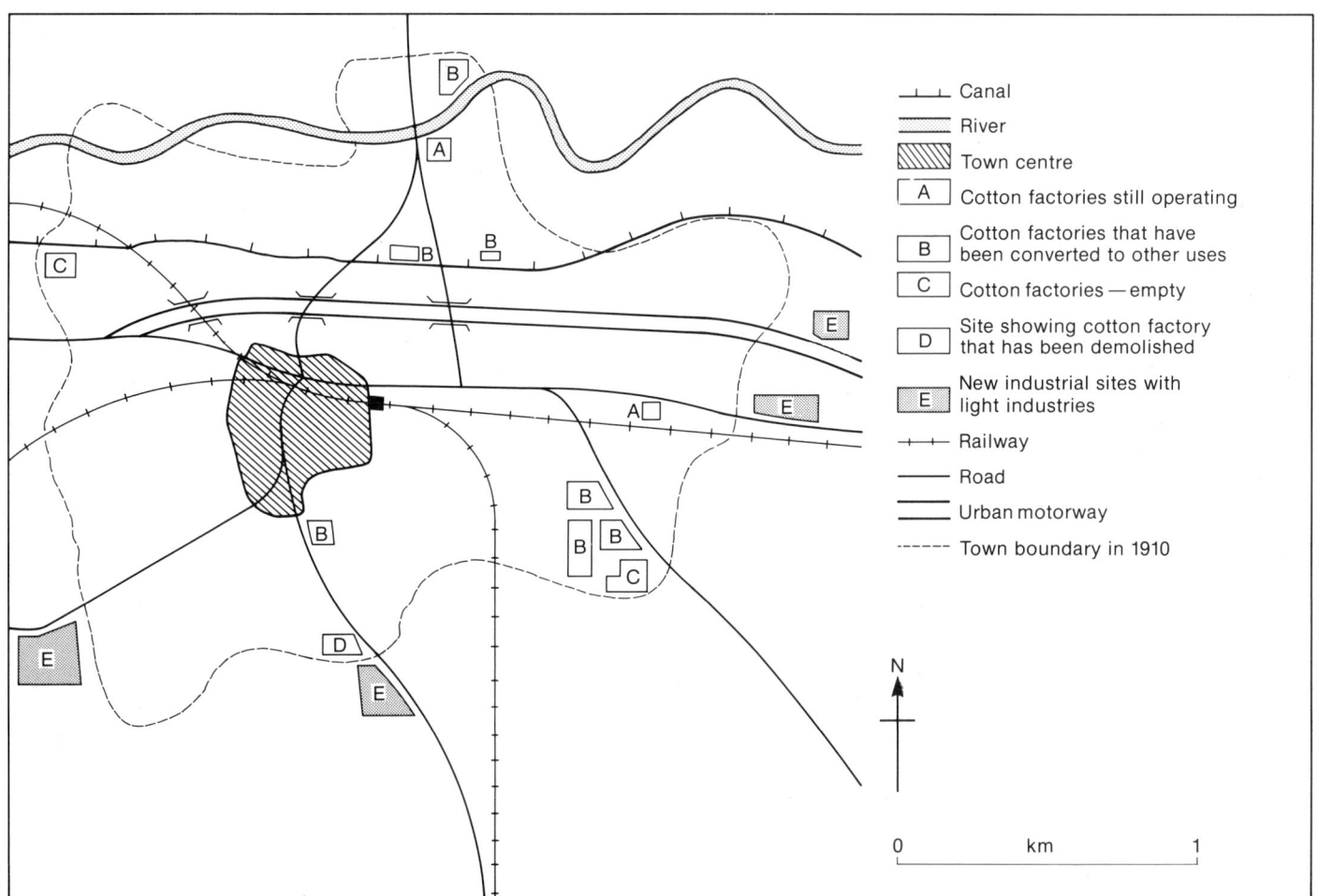

Figure 3.39. *A cotton textile town in northern England*
State three advantages for textile factories in this area. Count the factories in each group A, B, C, D. Draw a bar graph comparing the groups and describe what the graph shows.

Questions

1 Choose eight items of clothing you have and state what fabric is used in each. Compare your results with those of others and draw graphs or diagrams to show your findings.

2 What advantages do synthetic fibres have over natural fibres?

3 Why have factories making synthetic fibres been set up in areas like South Wales, Merseyside and Northern Ireland?

Shipbuilding

In the past, Britain has relied heavily on shipping. As an island, ships were needed to move both goods and people in and out of the country. Seventy years ago, more than two fifths of the world's ships were British. Many foreign vessels were also built in this country, so 80% of ships in operation were British built.

Shipbuilding is carried out in sheltered estuaries around the coasts. As large quantities of sheet steel are needed to make ships' hulls, shipyards are usually located close to main steel making areas. Examples are at Glasgow on the River Clyde, at Newcastle on the River Tyne and at Sunderland on the River Wear. These sites are also coal mining areas (the older iron and steel sites were on coalfields). In 1980, 75% of all British built ships came from yards in these three areas.

The total tonnage of ships built in British shipyards each year this century has not changed very much. But because ships have become larger, the actual number of vessels built has fallen since the 1950s. As a result, several shipyards have been forced to close down, while others have had to merge. In 1975, because many shipbuilders continued to lose money, the Government stepped in and nationalised the larger yards.

The Government has made one large company out of several smaller ones. It has provided money for the industry and now controls where ships are built. This prevents some shipyards from having too many ships to build while other yards lie idle. Such planning helps to reduce unemployment in the industry. It should also assist British shipbuilders to compete better with other countries.

There are fewer orders for new ships throughout the world. British shipyards have been particularly badly hit by competition from foreign shipbuilders. This is mainly because:
(a) production methods in Britain tend to be more outdated than in other countries, e.g. Japan.
(b) shipyards in Britain have often failed to finish building ships on time. Until recently poor labour relations were sometimes the cause of such delays.
(c) some other countries' shipyards are well subsidised by their governments, e.g. Japan. This gives these countries an advantage over British shipbuilders, who cannot build ships as cheaply. This is important as orders for new ships have dwindled since the 1970s and competition between shipbuilders is fierce.

Today, shipyards make a variety of vessels ranging from bulk carriers to container ships. They also produce warships, dredgers and North Sea oil platforms. Some yards specialise in the repair of ships.

Shipbuilding does not just mean building ships' steel hulls. There is a great range of jobs associated with the construction of ships. Painters, carpenters and electricians are all required to complete a ship. A group of workers will be responsible for making the engines.

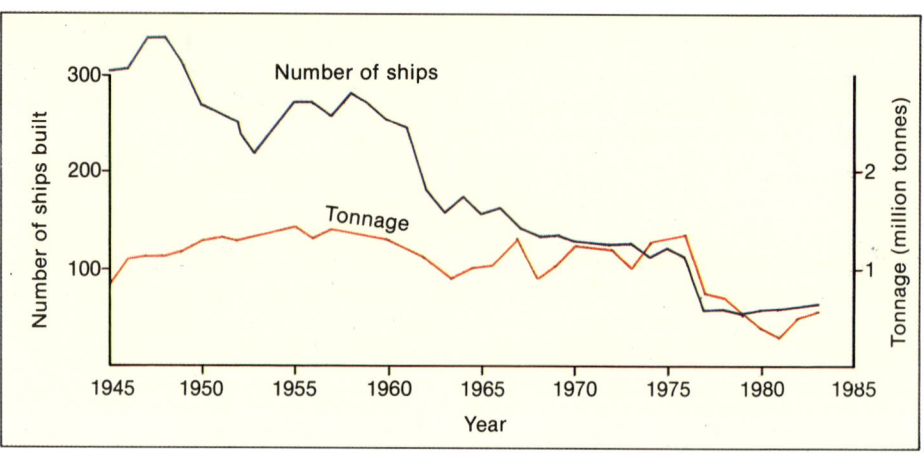

Figure 3.40. *Production of ships since 1945*
What do you notice about the number of ships and tonnage?

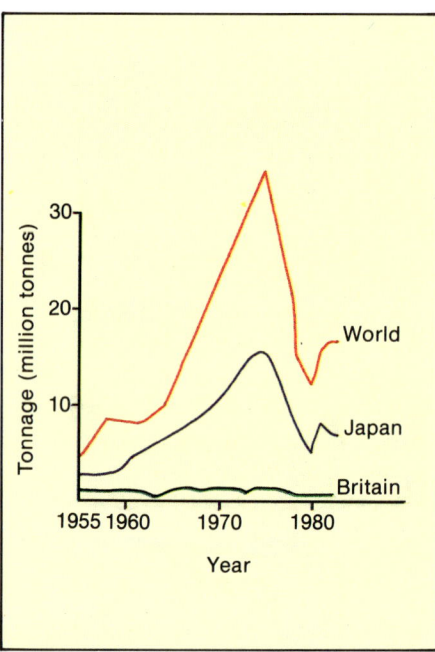

Figure 3.41. *Britain's shipbuilding industry compared with Japan and the rest of the world*
Describe the trends shown on the graph. Suggest reasons for these changes.

Questions

1 Describe and explain the main factors which have affected the location of shipyards. (Think of suitable possible sites, raw materials, workforce, Government help.)

2 List the different types of ships that are built in British shipyards.

Industry · SHIPBUILDING

Figure 3.42. *The main shipyards of Britain*

○ SHIPBUILDING ON THE TYNE

One region famous for its shipbuilding is the north-east of England. Two areas are important: the River Tyne and the River Wear. Newcastle is the main city on the River Tyne. It was between Newcastle and the coast that shipbuilding became important. Coal was discovered in this area and wooden ships had to be built to take it to London. Gradually, shipbuilders got other orders. By 1880, there were 12 shipyards on the Tyne. These provided a good market for large quantities of locally produced steel.

The Tyne was originally a twisting river. Later it was straightened and deepened to enable ships to be landed more easily. It was not long before much of the area between Newcastle and the coast was being used for shipbuilding. All the towns marked on Figure 3.43 had important shipyards.

In the mid-1950s this area produced 700 000 tonnes of shipping each year. This was half of the country's annual output of ships. By 1980 shipyards on the Tyne only produced 10% of the country's total output of 500 000 tonnes.

Only five shipyards remain in this area today. Wallsend is one of the most important. It is situated halfway between Newcastle and the sea. It covers 12 ha and employs 4000 workers. During recent years, the shipyard has been much improved and modernised. It now builds ships using prefabricated steel sheets. A wide range of merchant ships and naval ships is produced in the Wallsend shipyards.

Several firms on the Tyne have merged to try and cut costs. But even with some Government financial aid, further prospects for shipbuilding in the area do not look very bright.

Several thousand workers have already had to leave the shipbuilding industry. The Government has set up industrial estates to try and prevent very high unemployment. Team Valley, to the south of Gateshead, is one example. Just over 20 000 workers are engaged in a number of small industries, ranging from machine tools to electrical goods. A high proportion of the workforce once worked in the shipbuilding industry.

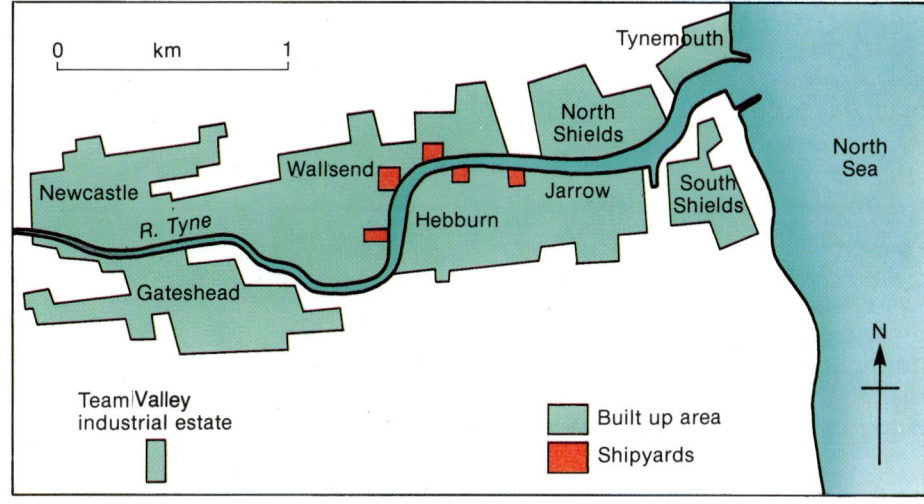

Figure 3.43. *Shipbuilding on the Tyne*

Figure 3.44. *A shipyard on the Tyne*
Why are covered factory buildings needed? Suggest what industrial activities might occur within them.

Light Industry

During the last 30 years there has been a rapid rise in the number of new, small industries. They are known as light industries. Most are grouped together on specially developed industrial estates. On these estates, there are factories of varying sizes producing a wide range of products. These vary from washing machines to packaging. Some companies do not actually make anything but are just distribution centres for products made elsewhere. Some factories are very small, only employing a few people. Others are much larger and employ hundreds of workers. In comparison with heavy industry, the range of materials and products made tends to be of a higher value but smaller in bulk. Some firms only make components which are then taken to be assembled elsewhere.

Firms are particularly attracted to industrial estates because they are often on the edge of towns. They offer room to expand and are away from congested, built up areas. Many of the factories offer pleasant working conditions which help to attract employees.

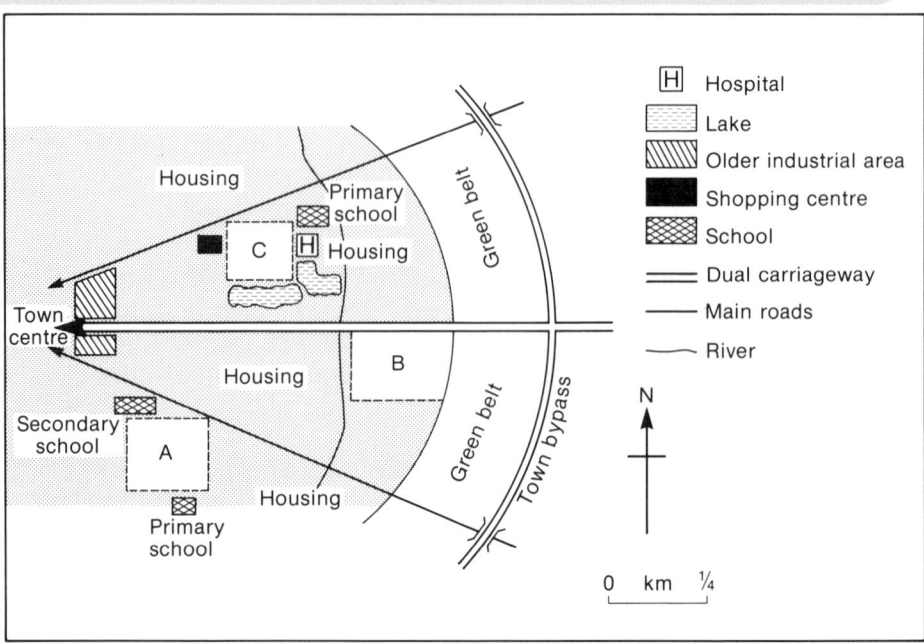

Figure 3.45. *Eastern part of a town*

Figure 3.46. *An industrial estate in Redditch*

Exercise

Figure 3.45 shows the eastern part of a town. Much of the area is made up of housing of varying ages. However three areas have been left and are now becoming wasteland. These are labelled A, B and C. The local council has decided to put these to good use. One is to be an industrial estate, another a recreational area with sports facilities and playground area, while the third is to be a landscaped open space with trees, walkways and grassy slopes.

(a) Suggest which areas the council might use for its industrial estate, its recreational area and its open space.
(b) Give reasons for your choice of the site for industrial development.

Map reading question

Look at Map 3 on page 125. An industrial estate is planned for this area. It will consist of 40 small units. Several sites have been suggested and these are: 053159, 040146, 060145, 067186. Locate each grid reference and list the advantages and disadvantages of each site. Which site would you choose?

Industry · LIGHT INDUSTRY

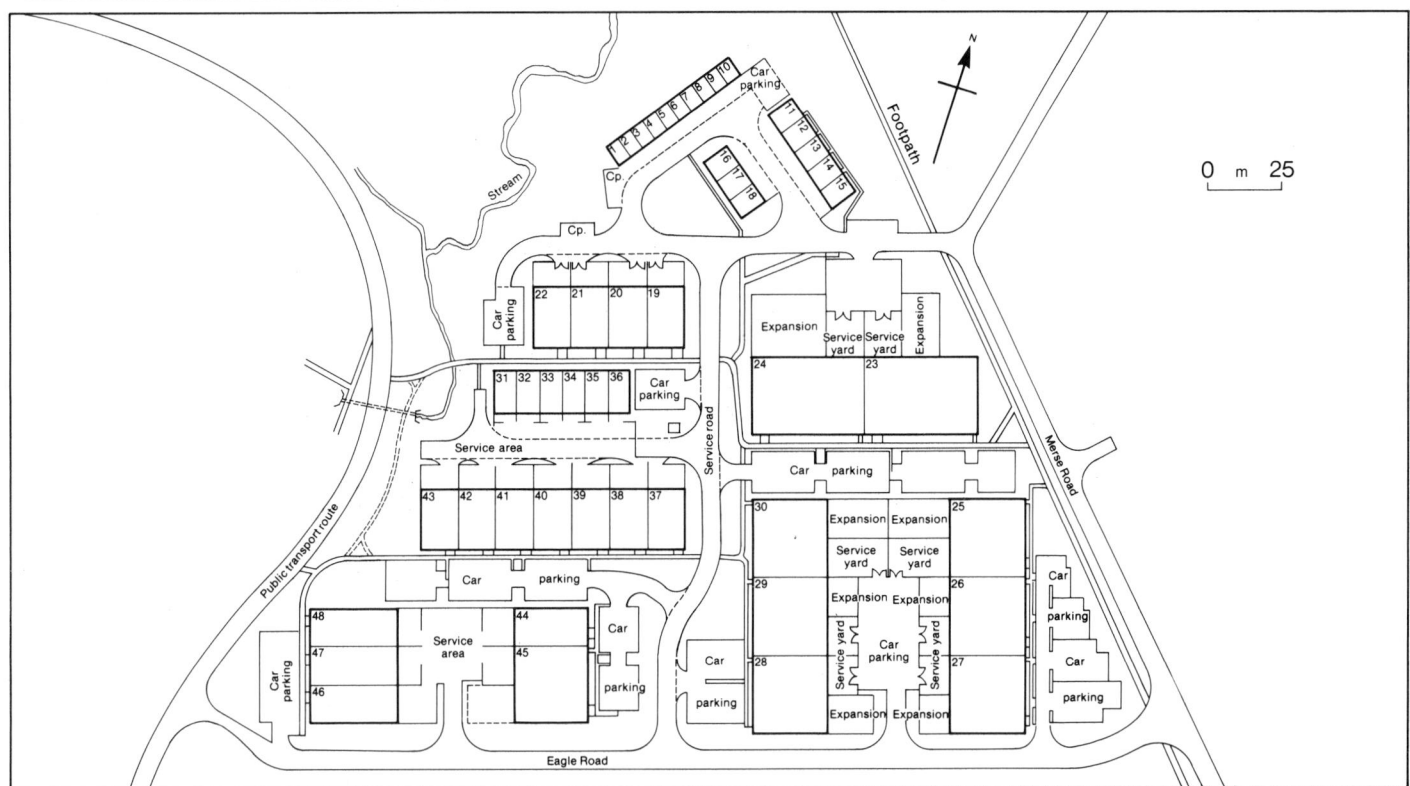

Figure 3.47. *Plan of an industrial estate at Redditch*

Figure 3.48. *Some of the industries occupying units on the industrial estate, and the areas they have come from*

Unit	Trade	Origin
1	Light industry	Local
2	Hi-fi turntables	Local
3	Warehousing/joinery	Birmingham
4	Chemicals	Local
5	Repair/sales of gold blocking	Worcester
6	TV aerials	Local
7	Instrument panels	Birmingham
8	Carpentry	Local
9	Gold embossing	Local
10	Light industry	Devon
11	Engineering	Local
12	Servicing and repair of lifts	Local
13	Light industry	New
14	Gig borers	Local
15	Pressed tool work	Local
16	Estate boards	Warwick
17	Empty	
18	Pickling and oiling metals	New
19/20	Cutting tools	West Germany
24	Making compressors	Local
25	Electronic equipment	Birmingham
26/27	Silver plate	Leamington
28	Exhibition contractors	Birmingham
44	Servicing of petrol pumps	USA
45	Plastics	Warwick

Questions

Study the photograph, map and table (Figures 3.46, 3.47 and 3.48) which show part of a recent industrial estate in the new town of Redditch in the West Midlands.

1 The map does not include some of the most recent factories. Which part of the photograph is not shown on the map?

2 Describe the layout of the industrial estate under the headings: road layout; provision for car parking and loading/unloading lorries; areas of grass and trees.

3 Describe the factories under the headings: size of buildings; shape; design and building material.

4 How does the area cater for firms of different size?

5 Draw a graph to show where the companies in the factories come from. Put the factories into five groups: local; Birmingham; elsewhere in Britain; overseas; new.
Write a short account to explain your graph.

○ DEVELOPMENT AREAS

Certain areas of Britain have a long tradition of one particular industry. The north-east of England and the River Clyde at Glasgow are famous for shipbuilding, and South Wales for coalmining and steel making. During recent years the demand for these products has changed. Fewer ships have been ordered from British shipyards. More profitable coalfields have been found in different parts of Britain, and steel can often be made more efficiently in new furnaces on coastal sites or imported from overseas.

As a result, many older industries are now in decline. So too are many smaller firms which exist to supply parts to these industries. In these areas there is now very high unemployment. When people are out of work, they do not have much money to spend. Many shops and small businesses have therefore declined. This has put even more people out of work.

In an attempt to solve some of the problems of industrial decline and unemployment, the Government has introduced a scheme to try and attract industry to towns like Corby in the East Midlands. They can now provide:
(a) financial help for new factories and machinery (up to 22% of total costs);
(b) financial help for new offices;
(c) grants to help skilled workers move to the area,
(d) grants for training workers in new methods,
(e) grants for changing from oil to coal fired boilers,
(f) help for firms wishing to produce new products,
(g) loans from the Common Market,
(h) preferential treatment, so that firms in these areas might expect to receive contracts from Government departments and nationalised industries, in preference to outside companies.

The areas which receive this assistance are shown in Figure 3.49.

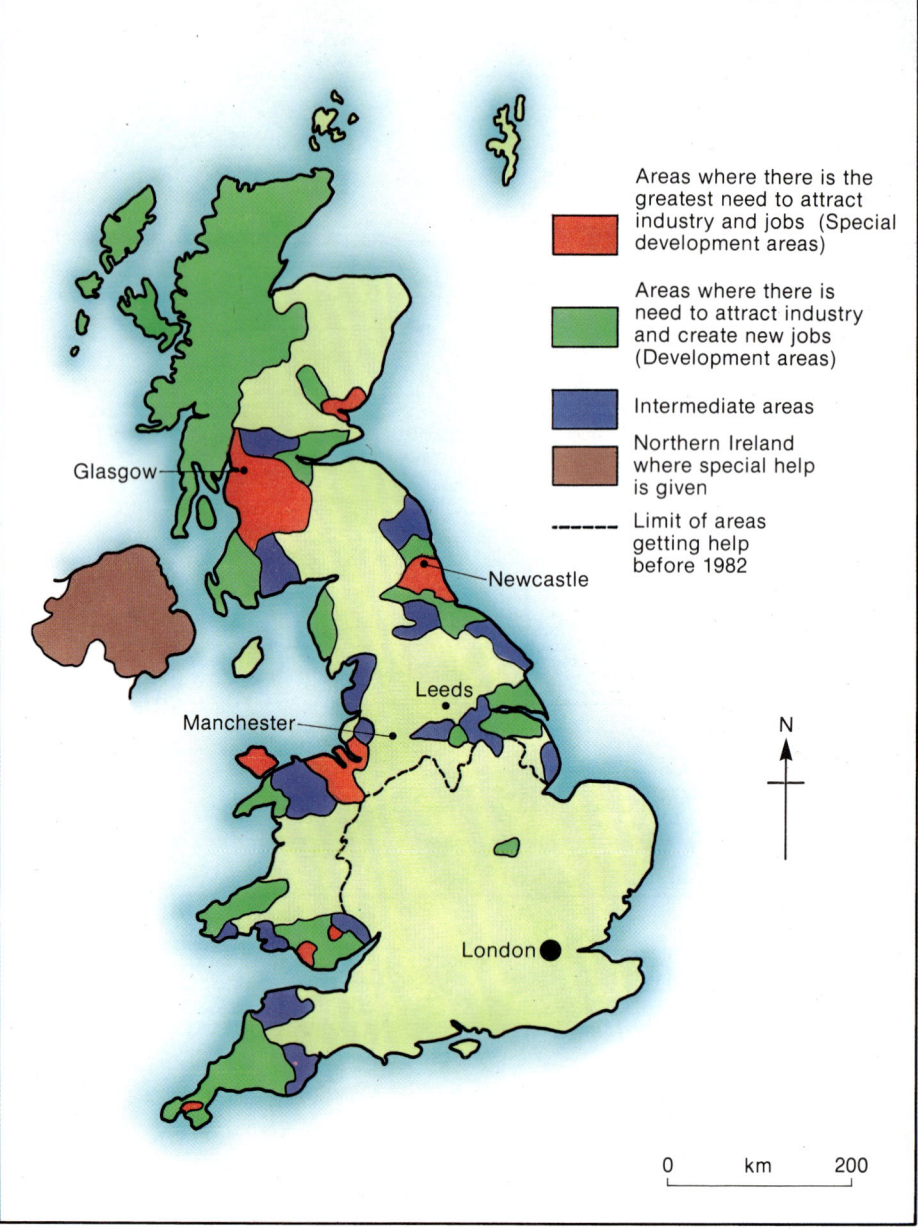

Figure 3.49. *Areas of Britain where expansion is being encouraged*
The Government is encouraging new industries and jobs in the areas shaded above.

Similar grants are provided in Northern Ireland, but more money is available there because of that area's special problems.

The Government has given special encouragement to tourism. Loans can be obtained for tourist craft industries, as well as for hotels. Particular assistance has been given to aid tourism in the north of Scotland.

Setting up these new industries can be a lengthy process. Planning permission has to be given to build factories and obtaining grants can take a lot of time. To try and speed up the process, new Enterprise Zones have been set up in certain parts of the country. Industries wishing to move to these areas are given as much help as possible, and as quickly as possible. The minimum of planning is required and decisions can be taken without delay. These areas are shown in Figure 3.50. In addition to Government schemes to attract industries to declining areas, the British Steel Corporation also provides sites for factories. These are on the sites of old steelworks.

Industry · LIGHT INDUSTRY

75

Figure 3.50. *Enterprise zones*
They were created in the early 1980s to try to make the introduction of industry into these areas as simple and as fast as possible.

Figure 3.51(a). *Unemployment in the East Midlands 1985*

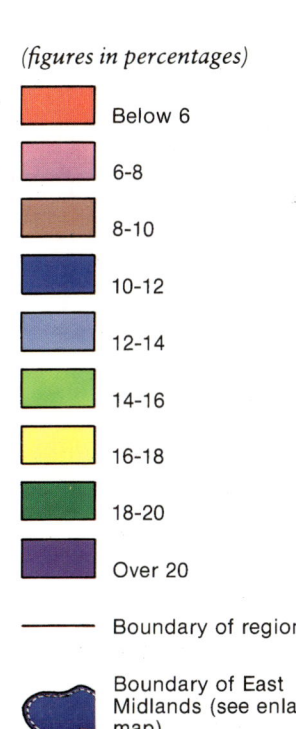

(figures in percentages)

- Below 6
- 6–8
- 8–10
- 10–12
- 12–14
- 14–16
- 16–18
- 18–20
- Over 20
- —— Boundary of region
- ┄┄ Boundary of East Midlands (see enlarged map)

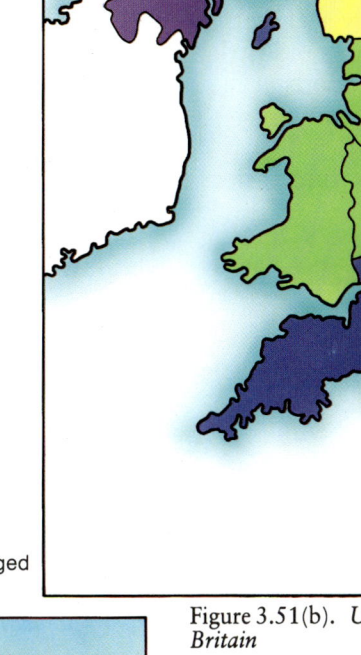

Figure 3.51(b). *Unemployment rates in Britain*

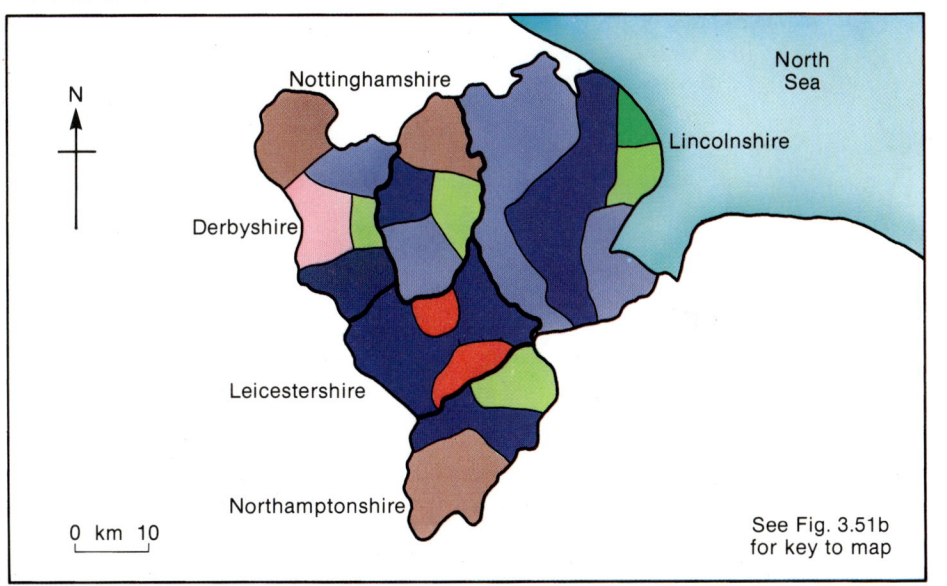

See Fig. 3.51b for key to map

Questions

1 Describe the areas where expansion is being encouraged.

2 Suggest reasons why these areas are getting help. Why doesn't the south-east need much assistance?

3 Describe where (a) the highest and (b) the lowest rates of unemployment are in Britain.

4 What effects does unemployment have? (Figure 3.53 will help you.)

Change – A North/South Divide?

○ THE NORTH

The main centres of Britain's heavy industry have always been in the north of Britain. During the nineteenth century, industries like shipbuilding prospered on the Rivers Tyne and Clyde. Thousands of tons of iron and steel were produced in Sheffield and other steelmaking towns. Yards of wool and cotton rolled off looms in Yorkshire and Lancashire. Coal was the fuel for the new factories, so all the industries grew up on and around northern coalfields. But by the beginning of this century, countries like Germany and the USA had copied Britain's industrial ideas. They produced goods in modern factories and shipyards, which competed with Britain's exports. Britain's heavy industries began to decline.

Apart from temporary recoveries during war years, the industrial decline has continued. Since the 1970s, all Britain's heavy manufacturing industries have been producing fewer and fewer goods. Look at Figure 3.52.

The results are deserted factories, depressed towns and very high levels of unemployment. Figure 3.51 shows how unemployment levels in the north of Britain compare with those in the rest of the country. There is little prospect that the situation in the north will improve rapidly in the near future.

○ THE SOUTH

The situation is rather different in the south of England. Many people work in light engineering, in clerical and professional jobs, and in the service industries. These areas of employment are expanding rather than declining. The new 'high tech' industries are developing mainly in the south. This part of the country has good transport links with London and Europe.

Figure 3.52. *Production figures for some of Britain's old established industries (thousand tonnes)*

Year	Pig iron	Crude steel	Ships	Cars (thousands)	Commercial vehicles (thousands)	Cotton yarn	Wool yarn
1960	12 032	18 938	1 400	283	784	360	244
1972	16 653	26 848	1 040	1 717	466	189	243
1984	8 388	13 704	525	888	296	95	131

COACHLOAD OF LIVERPOOL TEENAGERS SEE PRIME MINISTER

A coachload of teenagers arrived in London to meet the Prime Minister yesterday. They wanted to express their feelings about the lack of jobs in the area.

One teenager said she had tried for 30 jobs in the last 2 months without even an interview.

Another said, 'I don't think I'll ever get a job unless I moved right away from this area'.

SERIOUS PROBLEMS OF FINDING JOBS IN A YORKSHIRE TOWN

A survey of 30 former pupils who were in the same fifth form three years ago found that

- 1 had gone to a polytechnic
- 3 had gone to local colleges
- 4 had managed to find jobs with parents or relatives
- 2 had stayed on in the Sixth Form to get 'A' levels but their results were poor and they still don't have a job
- 3 went on secretarial courses and all have jobs
- 17 went on Youth Training Schemes. Only 3 got jobs afterwards, two gave up the scheme after six weeks. The rest are all looking for jobs.

THE JOBS ARE HERE BUT WHERE DO WE LIVE?

Some industries in southeast England are looking for workers. Firms are attracting people from the Midlands and the north of England with the prospect of a job and better wages. But it is not all good news. As George, a welder who moved from Liverpool to Tunbridge Wells in Kent, says:

'It's a good job with better wages than in Liverpool. But it's the people that get me – they don't talk to yer.

'It costs a lot to live down here. I was lookin' for a reasonable house so that my wife and children could have a new start. But I can't afford a house – they're three times the cost of those back in Liverpool.

'Wha's more – they don't know nothin' about hardship in the south. I never understood about the North/South divide till I arrived here.'

Figure 3.53. *Cuttings from local newspapers*

Industry · CHANGE – A NORTH/SOUTH DIVIDE

Workers in much of the south are able to find jobs more easily, and they are often paid higher wages. More people are able to afford to buy their own homes. Car ownership is also higher than in the north of Britain.

Figure 3.54(a). *Outline map of Britain, divided into 10 areas*

Questions

1. Study the table showing figures for some of Britain's industries (Figure 3.52).
   (a) State what has happened overall to these industries.
   (b) Choose one of the industries and explain why there has been this change.
   (c) Name two industries in Britain where the trend is the reverse of that in Figure 3.52.

2. Copy the outline map of Britain together with the boundaries of the areas (Figure 3.54(a)). Using the information from Figure 3.54(b) shade in the map to show the percentage of households with two cars. Remember the lower the figure, the lighter the shading. Select six different types of shadings, one for each of the following groups: under 8%, 8–10%, 10–12%, 12–14%, 14–16%, over 16%. Describe the pattern that you have produced.

3. Select one of the other columns in Figure 3.54(b) and do the same as you did in question 2. Choose your groupings and the shading you are going to use. Add a title and put in the key. Describe what your map shows.

4. In Figure 3.54(b), what other headings could be used to show the difference between North and South?

5. State briefly, in your own words, why there is such a difference between the North and South of Britain.

6. Read the newspaper cuttings (Figure 3.53) and then answer the following:
   (a) What advice would you give to the teenagers in the first passage?
   (b) Draw a bar graph to show what happened to the 30 fifth form pupils in the Yorkshire school. How do you think this compares with your own area?
   (c) Why doesn't George like the south-east of Britain?

7. What can be done to ease the problems of the North/South divide?

8. Will the economic differences between the north and south lead to greater social differences and more social problems?

9. What problems might someone from the north face if they moved to a new job in the south?

Figure 3.54(b)

Area	Unemployment (%)	Gross average weekly household income (£)	Percentage of households with two cars	Heating and lighting*	Food*	Transport (car, bus, train)*	Households without inside toilet (per thousand)*
North	17.6	150.14	6.8	7.18	25.83	17.24	1.4
Yorkshire and Humberside	13.7	152.89	9.7	7.49	26.17	14.74	1.6
North-west	15.4	169.80	13.4	7.67	27.22	18.40	2.0
East Midlands	11.6	163.10	12.4	7.36	26.54	18.38	1.8
West Midlands	14.7	166.31	14.6	7.59	27.47	18.18	1.9
East Anglia	9.6	166.39	15.4	8.25	28.70	20.59	1.4
South-east	9.2	196.62	15.0	7.70	28.95	22.08	0.8
South-west	10.6	166.87	16.6	8.62	25.80	19.25	0.9
Wales	15.2	161.36	12.1	8.70	28.18	19.05	2.3
Scotland	14.6	162.29	10.2	8.40	28.82	19.12	0.4
Northern Ireland	20.5	134.07	9.8	12.40	29.56	14.87	1.4

*Average weekly household spending (£) 1985.

SECTION 4 · Transport

During the last 300 years, there have been great changes in the methods and speed of transport. Initially, people walked from village to village or made longer journeys by horse, or horse and cart. Few people travelled very far. Some made uncomfortable journeys in stagecoaches, which carried the mail. Gradually, canals were built for barges to carry bulky loads, and the railways really began to speed up travel in the mid-nineteenth century. There were few cars on the roads eighty years ago. Today, rail and road travel have led to high speed trains and motorways. Concorde shows the progress that has been made in air travel in just over seventy years.

Roads

The first roads in Britain were built by the Romans. They were designed to help armies move around the country. The roads were carefully surveyed, and made out of stone cobbles. They were very straight, and linked towns and army camps right across England. Several modern roads follow old Roman courses. This can be seen easily on several Ordnance Survey maps. Figure 4.1 shows the courses of the main Roman roads.

After the Romans left Britain, these roads fell into disrepair. Unlike the soldiers, ordinary people did not need to travel very far from their villages. They only needed to go as far as outlying fields or nearby settlements to exchange goods. The tracks and paths they used for over 1000 years provided the basis for a large number of the roads we use today. It explains why so many of our present roads are very windy. They twist and turn through the same towns and villages as the old tracks used to, hundreds of years ago.

Although some early town streets were cobbled, most country roads were still rough tracks. In wet weather they often turned to seas of mud and were impossible to use. In the nineteenth century, as Britain became more industrialised, people wanted to carry goods around the country more easily. Engineers like McAdam began to build better roads. These roads needed a top layer of tar to make them waterproof. Modern roads are still covered in macadam (layers of broken stone).

Since the 1940s, there has been a huge increase in the number of cars and lorries using our roads. Most families now have the use of a car for work and pleasure. The increase is shown by Figure 4.2.

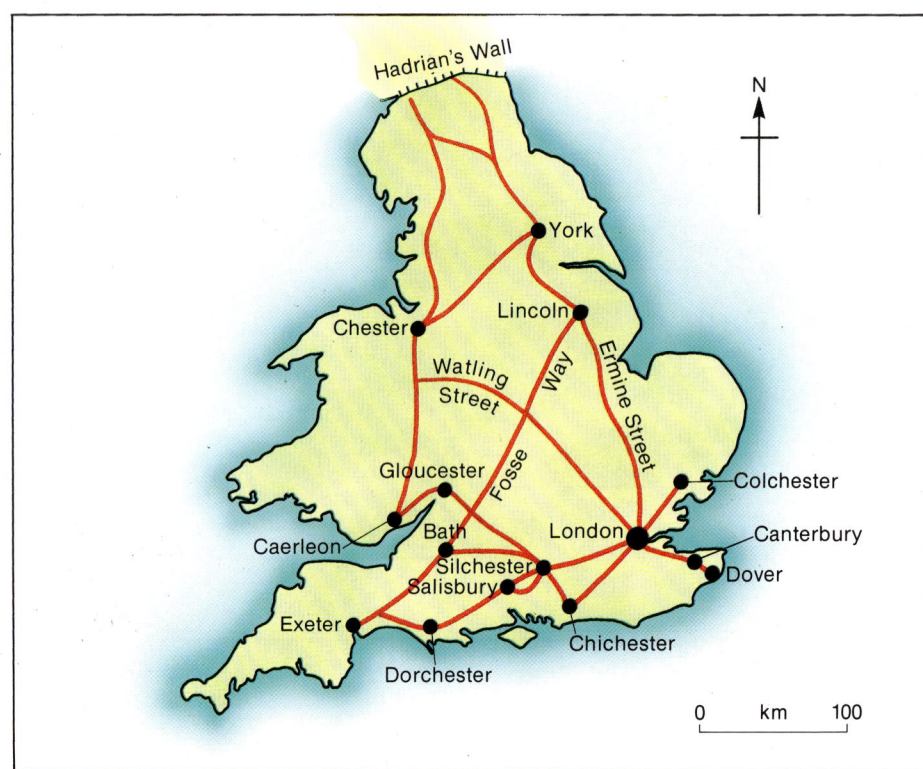

Figure 4.1. *Main roads in Roman times*
Also shown are the main towns. Describe the pattern of roads on the map. Study the map of the motorways in Figure 4.7. Is there any similarity with this map?

Figure 4.2. *The increasing numbers of cars*

1946	500 000
1950	4 million
1960	5.5 million
1970	10.5 million
1980	15.2 million

Unfortunately, many of our roads have not been improved for a long time. The increase in traffic

Transport · ROADS

Figure 4.3. *Piccadilly Circus, London. How do bus lanes help traffic problems in city centres?*

often causes great problems. There are frequent delays on narrow and windy main roads. Traffic can only move slowly where major routes pass through towns and villages. Congestion is often particularly bad in built-up areas where several major roads meet at busy junctions.

Heavy traffic in towns can be dangerous and cause serious pollution. Juggernauts passing through small streets can virtually shake buildings apart. Large lorries also make it difficult for pedestrians to move about safely. Exhaust fumes can rot bricks, causing the surfaces of buildings to crumble.

One simple solution is to build bypasses around towns or villages. A bypass (sometimes called a 'relief road') is a new road taking traffic around the outskirts of a town, avoiding the built up areas in the centre. If a bypass goes all round a town it is called a ring road. The city of Oxford is a good example. Oxford is also a route centre (i.e. roads come from many different directions). Today the city centre is almost free from through traffic, so there is less vibration and pollution.

Unfortunately, it is very expensive to build new bypasses and ring roads. They also use up valuable farmland. So many towns and villages are going to have to suffer pollution and congestion for some time yet.

Away from the towns, modern dual carriageways have reduced travelling times for both cars and lorries. Lorries can average about 72 km per hour along improved roads. This figure is reduced to about 40 km per hour along the older routes which twist through the countryside and built up areas. Lorries travelling along faster roads can make more journeys, and so help to make the transporting of goods cheaper.

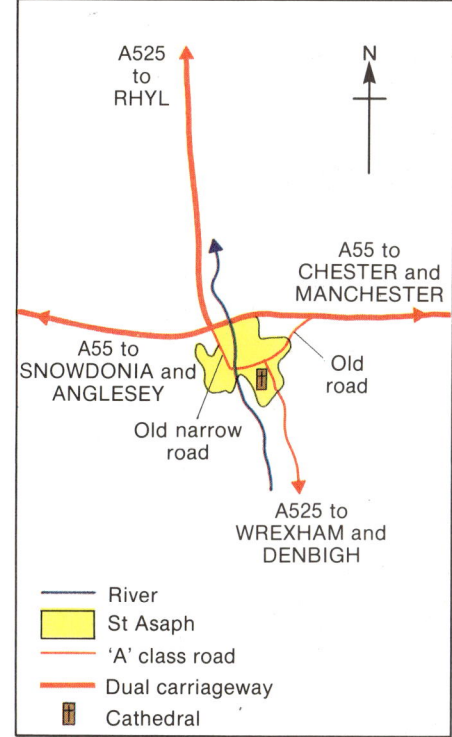

Figure 4.5 *Bypass at St Asaph*

Coaches also regularly use the roads. Since the late 1970s, coach travel has become more popular. Although it takes longer than the train, it is very much cheaper. There is a wide network of coach routes. Using fast roads and motorways, coach services are now offering serious competition to the railway. For example, the journey from London to Birmingham takes approximately 1½ hours by train, but only 2¼ hours by coach. The coach is much less expensive.

Road transport offers several advantages over other methods of travel. The most important are that:
(a) it is direct (you can travel from door to door);
(b) there are many different routes available;
(c) it is generally cheaper.

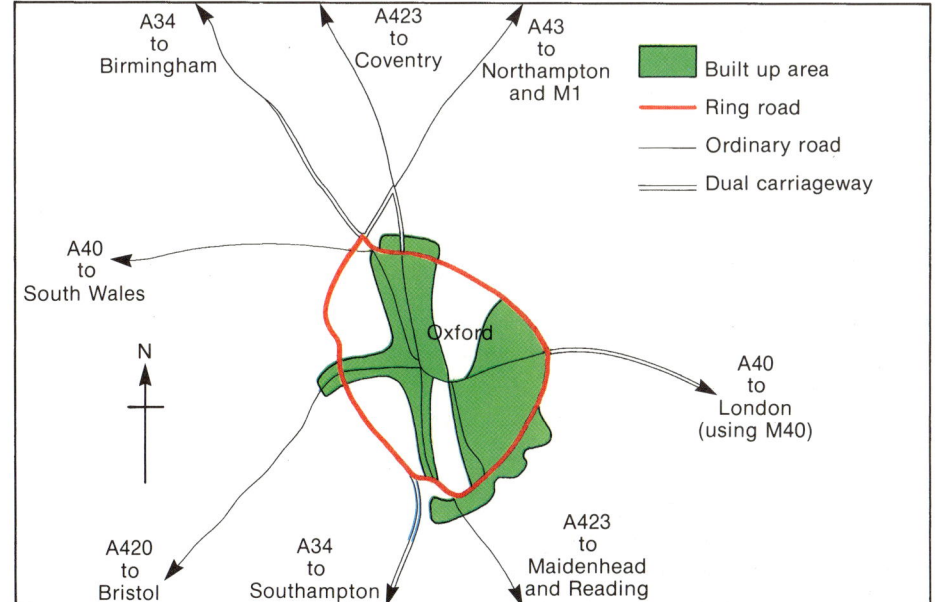

Figure 4.4. *Bypass around Oxford*
(a) How many roads used to go through the centre of Oxford?
(b) State why towns and cities like Oxford need a bypass.

The British Isles: Themes and Case Studies

○ MOTORWAYS

The number of cars and lorries using the roads rapidly increased in the 1950s. It became obvious that existing roads could not cope with all the extra traffic. It was decided to build a number of 'super roads' or motorways.

Figure 4.6. *Part of the M4 leading towards London*

Today, most of Britain's cities and industrial areas are linked by the motorway network. As mentioned, London is joined to Leeds by the M1; the M8 connects Edinburgh and Glasgow; Birmingham and Bristol are linked by the M5, which then continues to Exeter. This helps the holiday traffic heading for Cornwall. The M6 leaves the M1 near Rugby and passes close to the West Midlands on the way to Carlisle. The major cities in Lancashire and Yorkshire are joined by the M62. This is known as the Pennine Motorway and links Liverpool, Manchester, Leeds and Hull.

You will notice how many motorways lead towards London. They approach the city from all directions. Much of the traffic using these roads is travelling to London. But quite a proportion of the cars, lorries and coaches are only passing through the capital on the way to another destination. Lorries going to the continent from the north of England are one example. All this traffic causes severe traffic congestion in and around the centre of London.

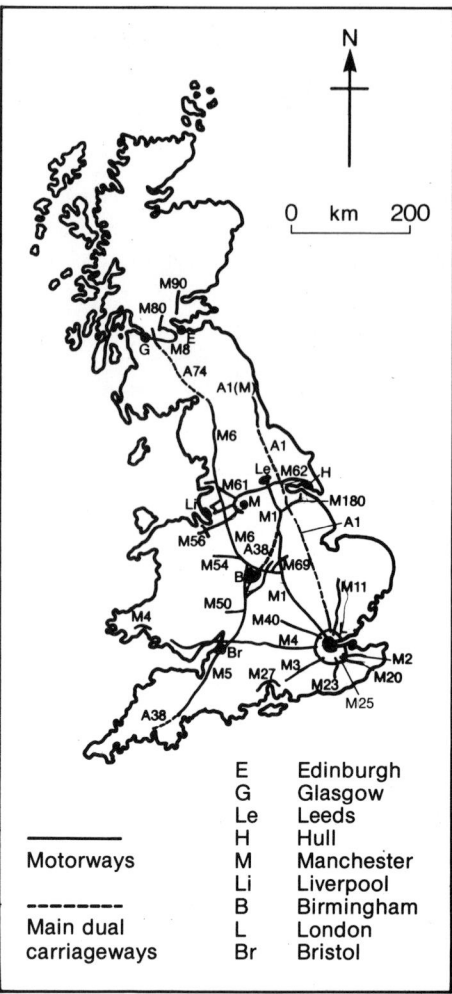

Figure 4.7(b). *The motorway system of Britain*

The motorways were intended primarily to link large cities like Leeds and London. They were built as straight as possible, with no roundabouts or traffic lights to cause hold-ups. Most were built with three lanes in each direction and only a few junctions.

The first stretch of motorway completed was part of the M6 around Preston in Lancashire. It was a success straightaway and traffic flowed around Preston very easily.

During the 1960s, motorways were constructed at great speed. Traffic quickly adopted the new routes. The M1 between London and Leeds is especially popular and it carries a very considerable number of cars and lorries.

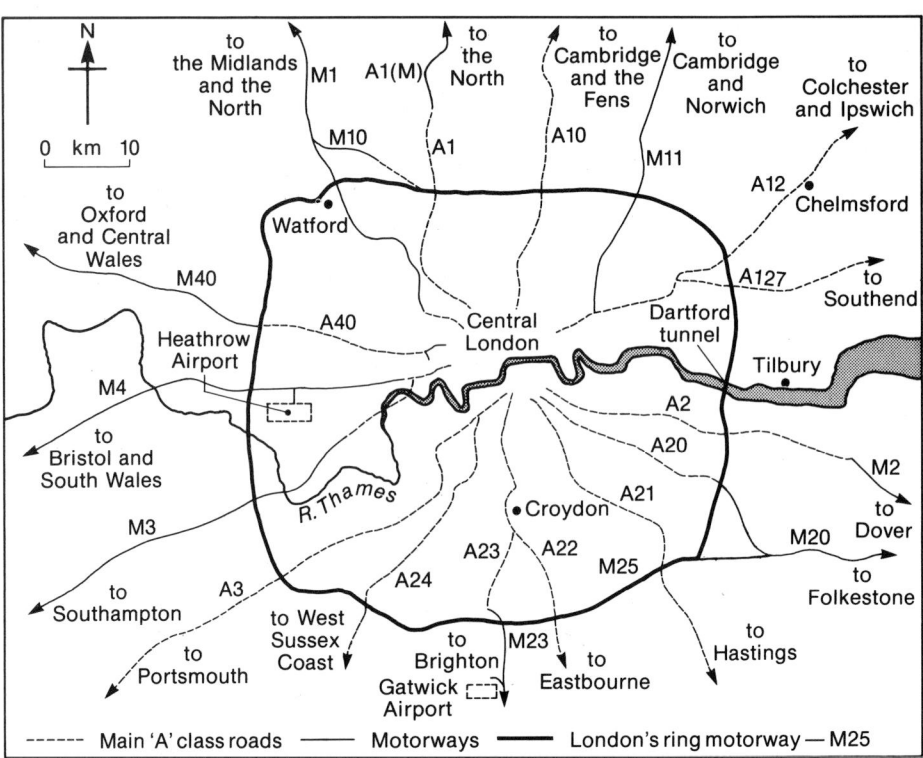

Figure 4.7(a). *Roads around London*

Transport · ROADS

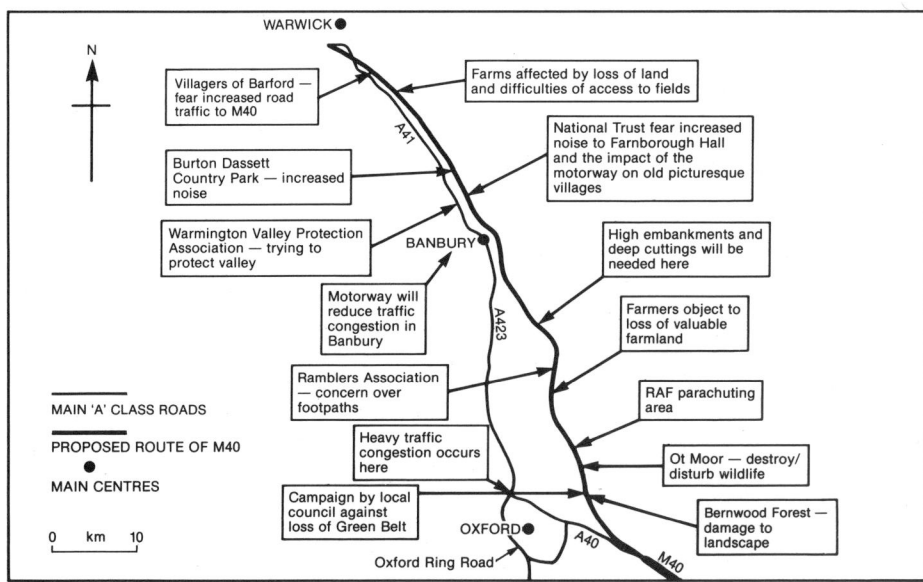

Figure 4.8. *The route of the M40 from Oxford to Warwick. The map shows some of the groups who would be affected by the motorway when it is built.*

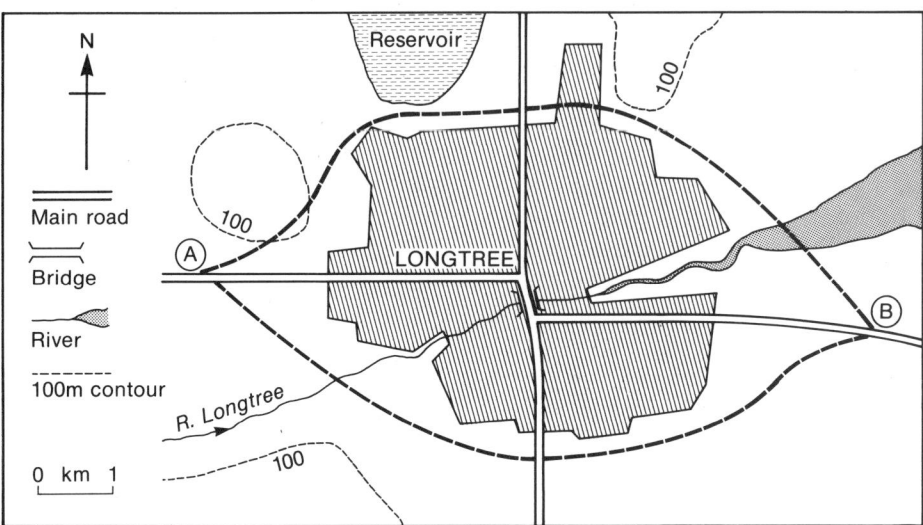

Figure 4.9. *A bypass for Longtree (two alternative routes from A to B)*

Cities like Manchester and Birmingham have built new roads to solve their traffic problems. Plans to build a new road around London were discussed for a very long time. After several public enquiries to decide the route, a new motorway around the capital was completed in the mid-1980s. It is the M25 – the London Orbital Motorway. The road was very expensive to build, and many people had objected that it had cut through farmland and countryside. Look at Figure 4.7a. It is hoped that the M25 will have eased problems in London. But are people using it?

Imagine you are travelling from Oxford to Dover. You will start on the M40 and end up on the M2 (or M20). You can either turn off on to the M25, or follow the A40 road into the centre of London. The A40 is a fast road, and much of the A2 out of London is a dual carriageway, but traffic conditions for about 15 km in the centre of the capital may be difficult. The M25 is a much longer route, though traffic should travel quite easily all the way on the motorway.

Which way would you go?

Another motorway is the extension of the M40 from near Oxford to Birmingham. One reason for the building of this motorway is to ease the number of cars, lorries and coaches using the M1 and M6 motorways between London and Birmingham.

Questions

1 Why is it necessary to build bypasses around towns?

2 What are the advantages and disadvantages of road transport?

3 Imagine a motorway was to be built across the map (Figure 4.9) from A to B to reduce the traffic flow through Longtree town centre. Copy the map into your book. Why is a bypass necessary for Longtree? Which route do you consider to be better? Give reasons for your choice and suggest why the other route is less good. Who might be likely to object to the building of a motorway and who might be in favour? Give reasons for your answer.

4 Suggest reasons why no direct motorway has been built from North Wales to South Wales.

5 One reason for the building of the M40 between Oxford and Birmingham is to ease the very high volume of traffic using the M1. Why is the M1 so heavily used? What problems does such a large volume of traffic cause on the M1? How will building the M40 help the traffic on the M1?

Map reading question

Look at Map 1 on page 123, and answer the following:
(a) Draw and identify five different types of road in grid square 4174.
(b) The motorway has been completed between 375724 and 465747. Suggest reasons why it was important to complete it.
(c) Why do motorway junctions take up such large areas of land?

Railways

○ PASSENGER SERVICES

Until 1956, almost all the trains in Britain ran on steam power. Services tended to be slow and infrequent. The engines were often old, dirty and unreliable. However, they used coal, which was plentiful in Britain.

In the late 1950s and early 1960s a modernisation programme was introduced. Steam engines were replaced by diesel or electric locomotives. New comfortable coaches were built and many of the smaller ones were replaced by longer carriages. By 1967, the steam engine had virtually disappeared.

Today, new modern trains link the main cities of Britain. They are known as 'InterCity' services. These trains are fast, clean, and comfortable, and run regularly. Coaches are air-conditioned and long distance services have refreshment facilities.

Two major advances have been made in InterCity services. The first started with the electrification of the main line linking London with Birmingham, Liverpool, Manchester and Glasgow. This was completed in 1974. Instead of three trains a day running between London and Glasgow, as in the 1950s, there are now eight in each direction. In the days of steam the journey took over seven hours. By the late 1970s, the travelling time had been reduced to five hours. This meant an average speed of 130 km per hour between the two cities. Look at the timetables below.

Figure 4.11. *A steam train in the early 1960s*

The Royal Scot

	London	Glasgow
1960	10.00	17.05
1986	10.45	15.45

Figure 4.10. *A High Speed Train*

Questions

1 Why was a modernisation programme needed for Britain's railways?

2 What improvements were made as a result of this programme?

3 What are the advantages of travelling by train, rather than by road between London and Glasgow?

Transport · RAILWAYS

The second major advance has been the development and design of High Speed Trains. These are capable of speeds up to 200 km per hour. These trains have been introduced on the London–Newcastle–Edinburgh, London–South Wales, London–Plymouth, and north-east to south-west routes. Once again, journey times have been reduced drastically. The trip from London to Edinburgh took only four hours 40 minutes in the mid-1980s.

The services discussed so far have been on InterCity routes. Around city areas and large towns trains are operated on suburban services. These bring workers and shoppers into city areas. The trains are shorter than those used on InterCity services.

The Tyne and Wear metro is an example of trains used for travel within city areas.

Railway time map of Britain

Maps are usually drawn to show the area and shape of land. However, Figure 4.14 has been drawn to show the time taken to travel along certain routes by rail. The shape of the country has been added afterwards as accurately as possible. The time taken to travel from London to Manchester, Leeds, Glasgow and Edinburgh is very short. Based on travelling times, these cities are close to London even though they are a long way away in miles. Journeys to towns and cities often much closer to London in miles can take a much longer time. Describe in your own words what the map shows. These maps are often produced to advertise new rail services. Why do you think this is done?

Figure 4.12. *The Tyne and Wear metro*

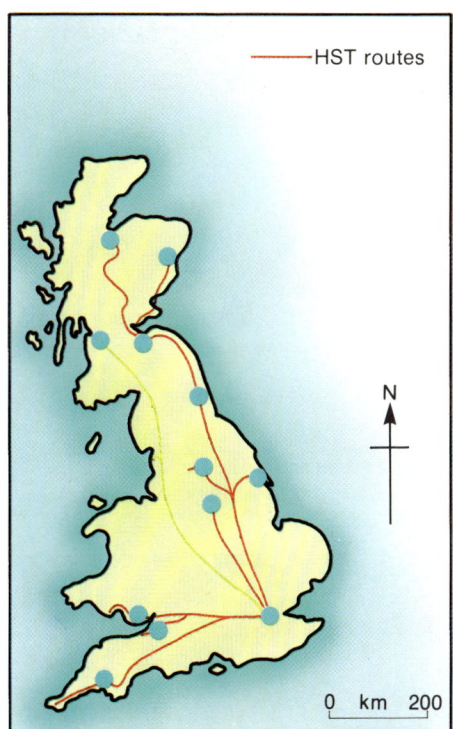

Figure 4.13. *The main High Speed Train routes*
These modern fast trains operate between many of the large cities in Britain. Are they concentrated on any one city in Britain? State, and give reasons for, one route along which High Speed Trains could operate in the future linking several important cities.

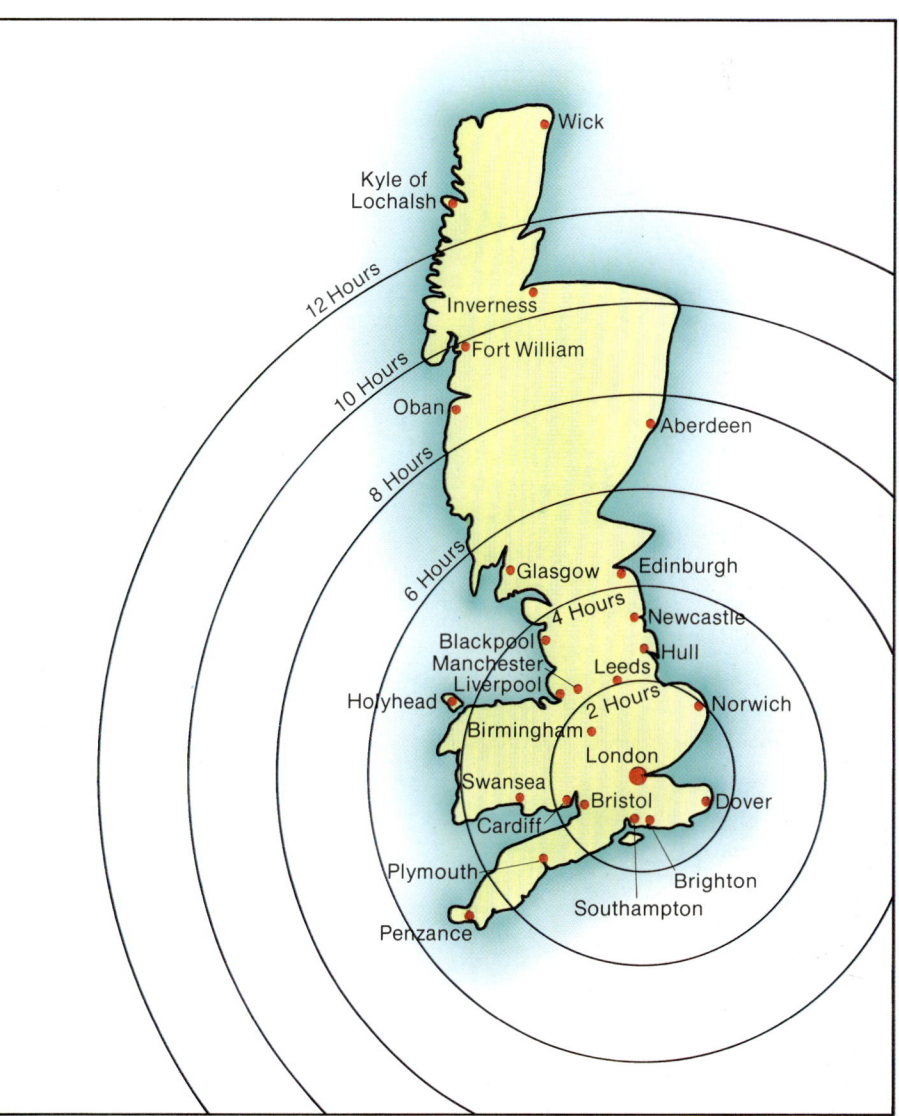

Figure 4.14. *A topological map*
Instead of drawing a map with the correct scale (like the ones in your atlas), this map is drawn based on the time taken to travel from London by the fastest trains. What advantage might a map like this have?

The British Isles: Themes and Case Studies

○ TOO MANY LINES

Study the map of railways in the West Midlands in Figure 4.17. The left hand map shows the railway network around Birmingham in the late 1950s. Many lines carrying passenger trains were uneconomical to run. Some trains operated with only a very few passengers. In places, there were also two different routes running side by side (between Birmingham and Wolverhampton, for example). This could not continue as it was very expensive to maintain track to a high standard, and costly to provide so many trains. A plan was introduced to close uneconomical lines and concentrate trains on more profitable, selective routes. This became known as the Beeching Plan. The right hand map shows the network that remains.

In other areas, many disused lines and railway buildings have been put to other uses. Look at Figures 4.15, 4.16, 4.18 and 4.19.

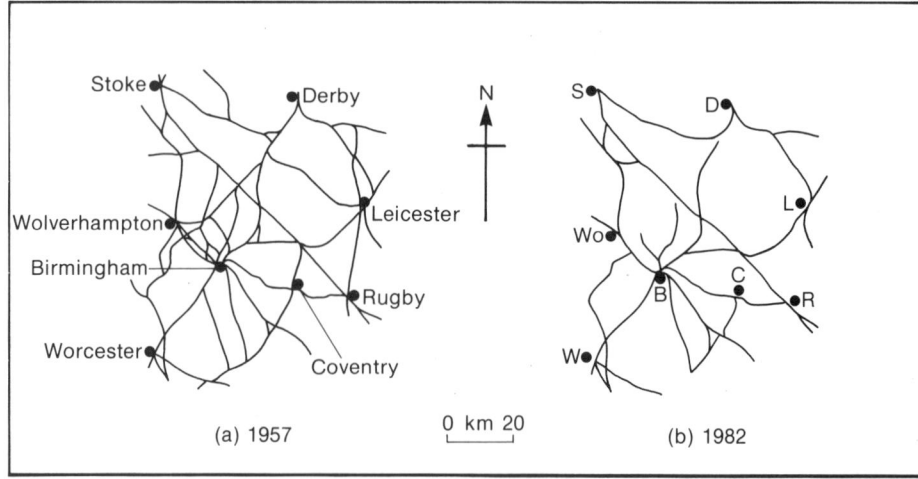

Figure 4.17. *Passenger rail services – West Midlands*

Figure 4.15. *The Bluebell Railway in Sussex now runs steam trains for pleasure trips.*

Figure 4.16 *(left) a disused train and railway building in N. Yorkshire has been turned into a hotel and restaurant.*

Figures 4.18 and 4.19 *(above right and below) show a disused station in Somerset, before being taken over by rail enthusiasts. What could be done with a station like this? (See Figure 5.39 on p. 115 for another idea.)*

UNECONOMICAL LINES REMAIN

Wales represents an area where a large number of railway lines were forced to close. Before closure took place, an alternative form of transport had to be provided. This was usually a bus service. However, in some areas of Wales, uneconomic lines had to be retained. It was too expensive to provide an alternative form of transport. In addition, some lines were reprieved on social grounds.

The Central Wales line from Shrewsbury to Swansea is a good example of a line kept open for social reasons. Situated between these towns are several small market towns, such as Llandrindod Wells, Knighton and Llandovery. There are also many small isolated villages away from main roads. To close this line would mean hardship to people in both the villages and the towns. Travelling by bus would mean a slow, laborious journey taking much longer than the train. Therefore, the Government pays a large subsidy to keep the line open. In return, there have had to be economies. For example, many of the stations have been turned into halts, which do not have any railway staff. The train operates rather like a bus, with the guard issuing tickets to passengers on the train.

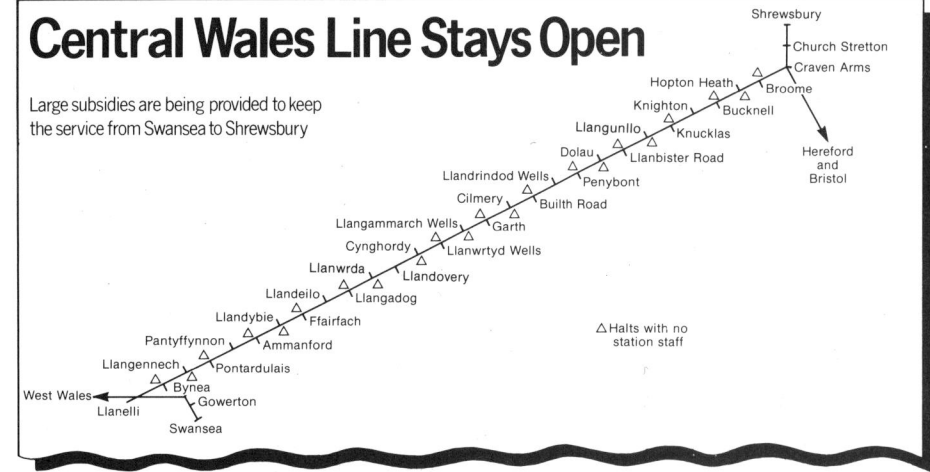

Figure 4.21. *The Shrewsbury–Swansea line*
Locate the line on your atlas. How many stations are there between Shrewsbury and Swansea (a) with staff and (b) with no staff?

Questions

The maps in Figure 4.22 show the network of passenger railways in Wales, for 1959 and 1982.

1 Draw a map of Wales and mark on the railway lines in use in 1982.

2 Using an atlas, name and locate the following towns: Cardiff, Newport, Fishguard, Swansea, Holyhead, Llandudno, Aberystwyth, Chester, and Pwllheli.

3 What happened to the network of passenger lines between 1959 and 1982?

4 Which areas suffered most closures?

5 Why did the two lines marked X and Y remain open?

6 The line from Shrewsbury to Swansea remained open on social grounds. Suggest another line in Wales that has remained open for the same reasons. Look for a line where there are few large towns.

7 Referring to the atlas, name three sizeable towns which no longer have a rail service.

8 Look at the atlas map of northern Scotland. Describe the pattern of railways in this area and explain why some of these lines have been kept open.

Figure 4.20. *The Shrewsbury to Swansea, central Wales line*

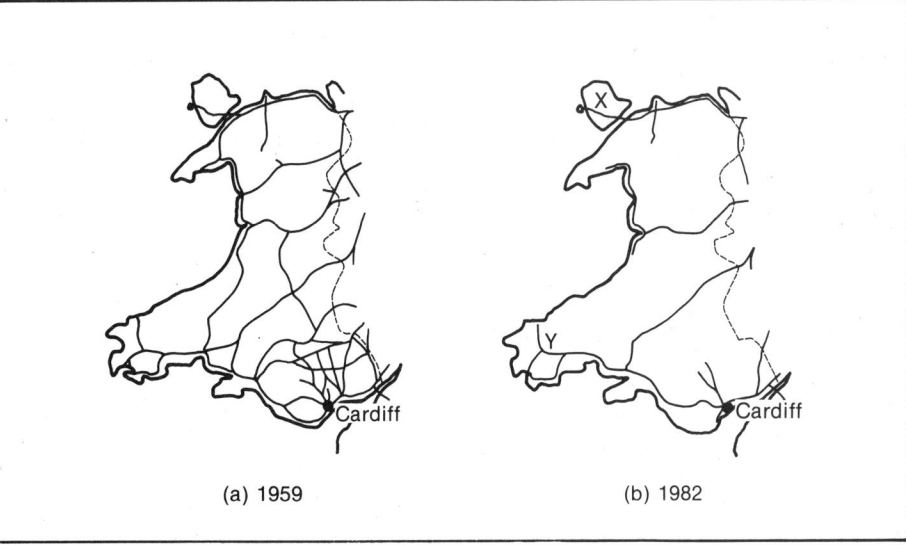

Figure 4.22. *Passenger lines in Wales*

FREIGHT TRAINS

In addition to passengers, the railways carry freight. In recent years, the amount of goods carried by rail has declined. Much of the 'lost' traffic has gone to road transport. Today, 16% of freight is carried by rail, compared with 67% by road (see Figure 4.23). For distances less than 150 km, it is cheaper for goods to be transported by road, taking them directly from door to door. Rail transport is more competitive with road haulage for distances over 150 km.

British Rail has concentrated on four main types of freight train.

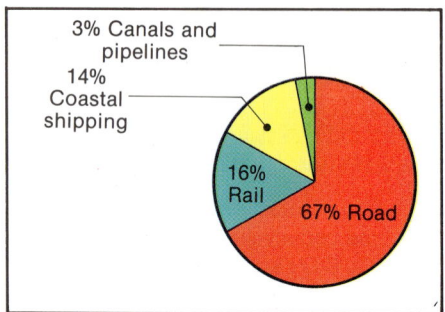

Figure 4.23. *Freight carried in Britain by different forms of transport*

Merry-go-round trains

Special trains transport bulk cargoes, such as iron and steel. They operate continuously without stopping to load and unload, as shown in Figure 4.24. Trains are made up of large hopper wagons which can be filled as they pass under the loading plant. Each truck can be filled in less than a minute, as the train travels at less than 1 km/h. The train then continues its journey on the main line and travels to its destination. When it arrives, the trucks pass over the discharge bunkers. The bottom of each hopper is opened, and the load falls into spaces beneath the track.

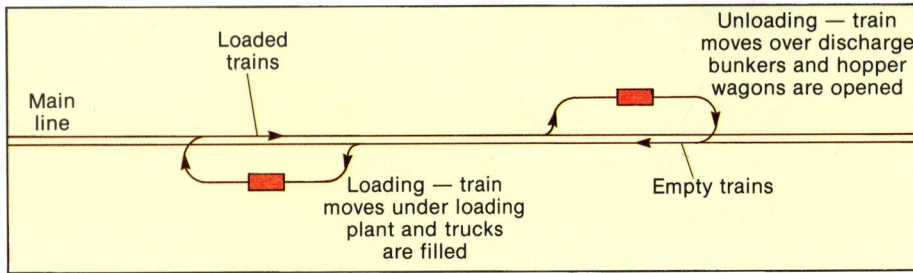

Figure 4.24. *Merry-go-round system*

Figure 4.25. *Number of merry-go-round trains per day*

	1967	1972	1977	1982
Number of trains	35	148	195	300

Merry-go-round trains are becoming increasingly popular, as is shown by Figure 4.25.

The National Coal Board makes considerable use of the merry-go-round system. Trains take coal from collieries and transport it to coal-fired power stations. West Yorkshire provides a good example of this (see Figure 4.26). When the Selby Coalfield is in full operation, coal from this area will be used in the power stations. The trains will need to use the existing track through Selby. However, by that time, high speed trains will be using the line too. It is proposed to build a new section of track around the town for the faster trains. This will avoid slow merry-go-round trains and very fast high speed trains having to use the same line.

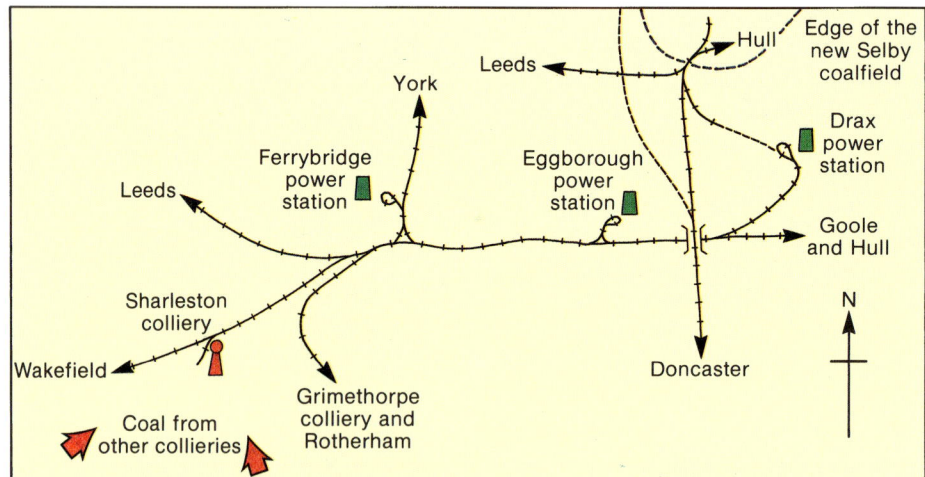

Figure 4.26. *Merry-go-round routes of West Yorkshire*

Figure 4.27. *A merry-go-round coal train*

Freightliners

These trains carry containers of regular size and shape. These can be loaded on to lorries, railway trucks or ships with the minimum of effort. Freightliner terminals have been set up in all major cities and ports.

Transport · RAILWAYS   87

Figure 4.28. *A freightliner*

Speedlink
A high speed selective wagonload service has been developed to handle smaller quantities of goods. One wagon can be loaded with a particular type of commodity. It can then be transported as part of a fast freight train, from one centre to another. This is a new service, with about 100 trains per day carrying eight million tonnes of freight per year. The network in operation by the mid-1980s is shown in Figure 4.29.

Company Trains
Private firms needing to transport large consignments of goods regularly can use special vans. Ford, for example, have a regular service between their factories at Dagenham and Halewood. Two of these trains operate each day between the two factories. The English China Clay Company operates a nightly service, taking china clay from St Austell to Stoke-on-Trent for the pottery/ceramics industry.

Questions

1 What are the advantages of using; (a) merry-go-round trains; (b) container trains?

2 Over a distance of 150 km, rail transport is more competitive for freight than road transport. Suggest reasons for this statement.

3 Many people believe that there are too many heavy lorries on our roads. Suggest ways in which more freight could be attracted to the railways.

Figure 4.29. *Speedlink services*
The main routes are marked on the map. The more lines between two stations, the more important is the route. Describe the pattern of the network of routes.

Ports

Britain's population has increased greatly over the last 200 years. More people than ever now need feeding. As Britain can no longer grow all the food we require, much has to be imported from overseas. Examples include wheat from Canada and sugar from the West Indies. We also have to import other goods like timber and minerals for our industries. To pay for all the imports, Britain has to export products such as machinery and chemicals. This 'trading' is important to the country.

Britain is completely surrounded by water, so nearly all imports and exports have to be carried by sea. Only the more valuable, less bulky products can be transported by air. Today's ships are very large compared with the ships of a hundred years ago. Ports have to accommodate these large ships, and so need deep water. Ships also have to be protected from strong winds and sea currents when loading and unloading. For this reason, most ports have grown up along sheltered estuaries. If necessary, the estuaries can be dredged. The surrounding land is also usually flat and firm. This enables docks to be built, and takes the weight of the quayside and storage areas. Figure 4.30 shows the location of Britain's major ports.

All ports need to have a large area where goods can be delivered and collected. This area is known as a hinterland. London's hinterland includes much of south-east England, the area of the Thames Basin, and extends as far north as the Midlands. With improved road

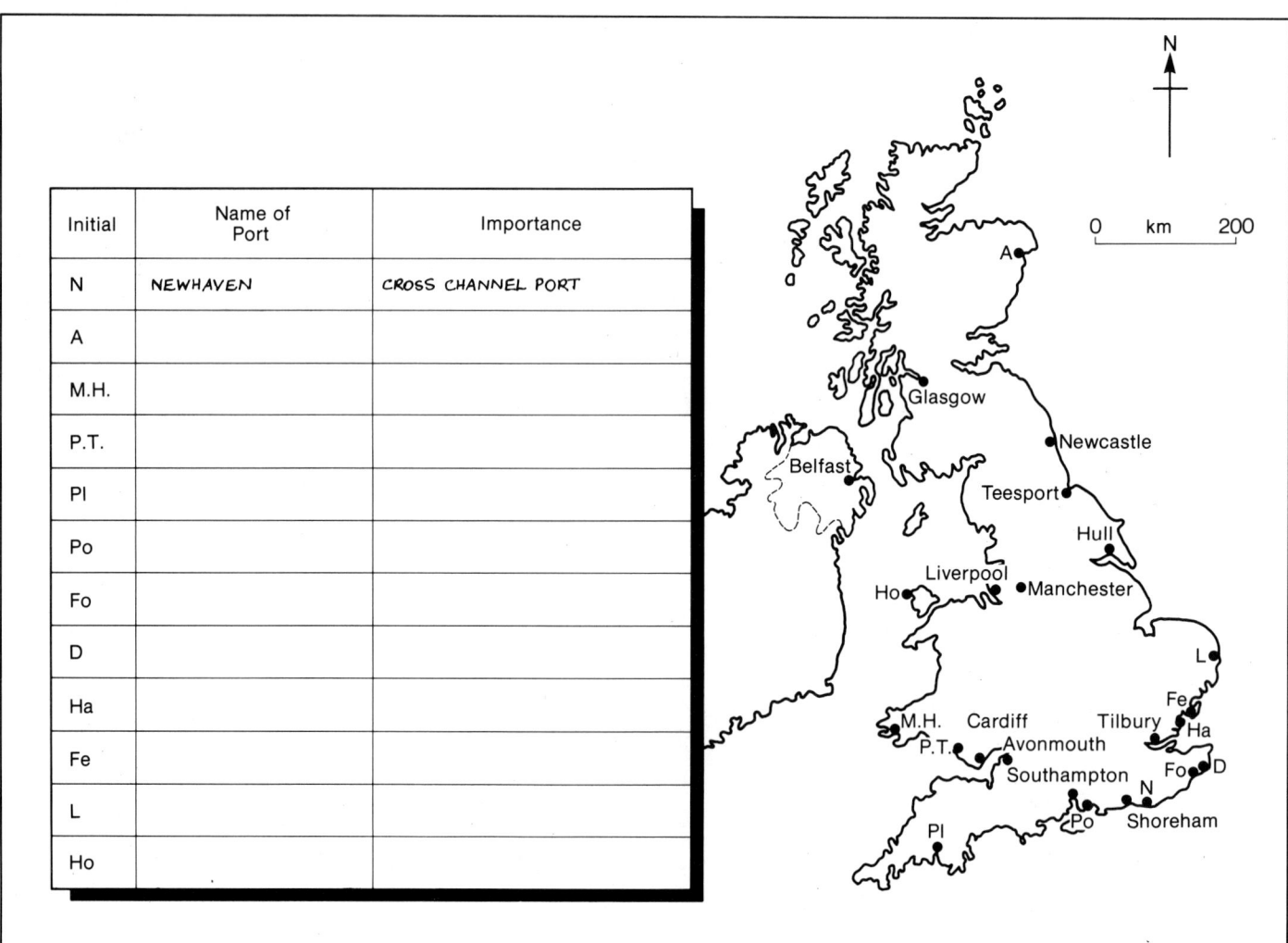

Figure 4.30. *Some important British ports*
(a) Copy the ports onto an outline map of Britain. Copy the table shown.
(b) Using an atlas, name the ports and state their importance. *(There are two naval ports, three cross channel ports, two container ports, a fishing port, an iron ore terminal and two oil terminals with fishing. The first port has been named.)*
(c) Explain why Manchester has been marked as a port. *(Use an atlas to help you.)*
(d) Explain why there are few ports in Scotland.

Figure 4.31. Containers being unloaded at Liverpool
(a) Describe how the containers are being unloaded from the hold of the ship.
(b) How many people might be involved in unloading this ship and what might they do?
(c) What are the advantages in having containers all the same size?

and rail communications, the hinterland of ports has grown. For example, goods to and from Birmingham may use any one of eight ports quite easily.

Some ports have tended to specialise in handling one commodity. For example, Milford Haven handles petroleum and Port Talbot handles iron ore. (Both ports are in South Wales.) Other ports like Felixstowe specialise in containers (see Figure 4.31). Most of the larger ports, such as London, Bristol and Liverpool have maintained a mixture of goods.

The ships carrying such products are also changing. It is cheaper to transport goods in large quantities so ships have tended to grow in size.

Figure 4.33. A Shell tanker carrying petroleum

ULCCs (Ultra Large Crude Carriers) and smaller tankers bring petroleum to our deep water ports. Bulk carriers transport solids such as iron ore and grain. Grain is sucked from the holds of the ships and tipped on to conveyor belts. Up to 2000 tonnes can be moved each hour in this way.

Roll-on roll-off ships are used for carrying containers, lorries and cars. Part of one end or both ends can drop down, so that vehicles can be driven straight on and off the ships. Some docks have declined because they are unable to handle these newer types of ships and the methods of loading and unloading. The old docks of London have closed down for this reason.

Figure 4.32. A passenger ferry. These ferries also use Britain's ports.

PORT OF LONDON

A port has existed at London since Roman times. It then handled exports such as skins, corn, metals and iron. It handled imports such as glassware, jewellery and pottery.

The port developed for the following important reasons:
(a) it was a long sheltered estuary that provided protection from the North Sea;
(b) the town of London was an administrative centre and offered trade;
(c) it allowed easy access to some of the big European ports.

Figure 4.35. *Tilbury Docks*

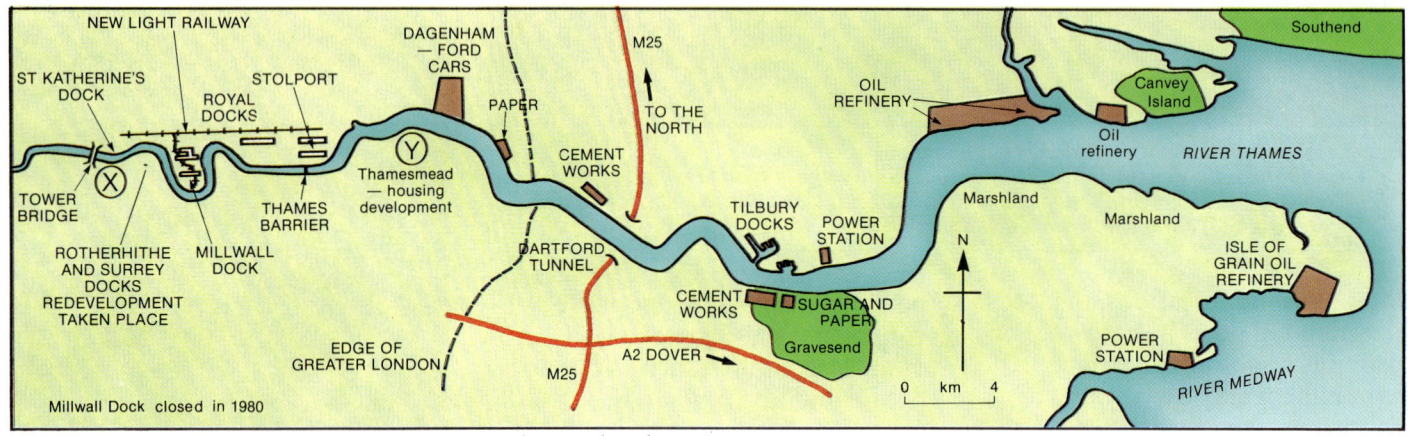

Figure 4.34. *The Thames Estuary, its main port and riverside industries*

The Port of London grew rapidly in importance during the eighteenth and nineteenth centuries as trade with other countries increased. The present docks (with the exception of Tilbury) were built between 1802 and 1890. A number obtained their names from the areas of the world with which they traded, e.g. the Indian Docks. Goods were imported from all over the world. For example, wool was brought from Australia, bananas and sugar from the West Indies, timber from Canada and Scandinavia, and tea from India.

As London grew in importance, so did the port. However, in the 1950s the port started to decline for the following reasons:
(a) the very rapid rise of some continental ports, such as Rotterdam in the Netherlands and Antwerp in Belgium. Both of these competed with London for world trade;
(b) the old docks had little room for expansion, because the immediate areas around were built up with housing;
(c) the methods of loading and unloading were out of date. Little money had been spent on improvement. Labour relations were not very good. It became expensive to use the port. Other ports near London began to attract some of London's trade (e.g. Felixstowe to the north-east).
(d) ships were getting bigger and had problems navigating up the River Thames.

With the decline in traffic, 2 ha of docks closed. St Katherine's Dock became a yacht harbour, as it was located close to the Tower of London and Tower Bridge. Both London Docks on the north bank, and Surrey Commercial Docks on the south side have closed and are awaiting redevelopment.

TILBURY DOCKS

So far, little has been said about docks downstream from London, at Tilbury. These were opened in 1886. Major changes took place in 1963 when a new deep water dock extension was built. It now has container handling facilities. Modern equipment speedily transfers containers to and from ships.

The port can compete now with Rotterdam and Antwerp. It can handle the largest ships that are operating. There is room for expansion, warehousing and storage. It is slowly starting to recapture some of the trade lost to ports like Felixstowe.

Today, Tilbury imports include about a third of all Britain's logs, timber, pulp and paper. These products are imported from Canada, Scandinavia and Africa, often using purpose-built ships. Such vessels can carry 25 000

Transport · PORTS

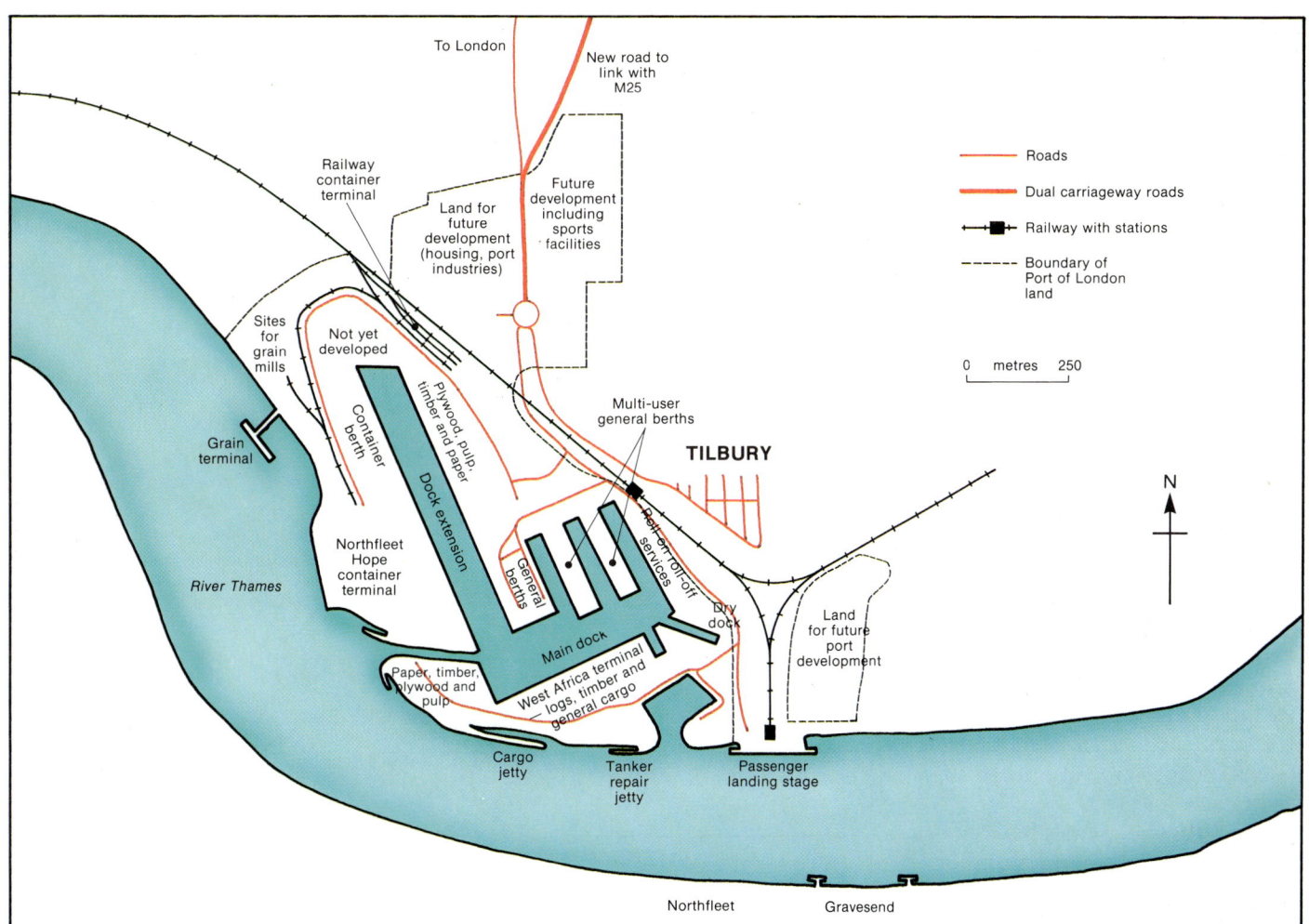

Figure 4.36. Tilbury docks
(a) Study the map of the docks and try to identify each one in the photograph.
(b) What are the advantages of this site for a dock?
(c) Describe (i) the layout of the docks, and (ii) the shape and size of the storage areas and warehouses.

tonnes of timber products. Alongside the docks are specially built warehouses and storage areas which enable the timber to be transported easily to its destinations. There is a good market in London for such products as newsprint, stationery and office furniture, as well as building materials.

Containers are an increasingly important part of Tilbury's trade. The recently built Northfleet Hope Container Terminal can handle the largest container vessels in the world. Facilities exist for loading containers on to lorries for distribution around the country. However, 20% of containers travel by rail, and a new container terminal has been completed for even faster distribution.

Grain is imported into Tilbury. Twenty per cent of all Britain's imported grain comes in through the port. Two thousand tonnes an hour can be transferred from bulk container ships. Large storage silos have been built to hold the grain until it is collected.

All the latest types of ships can be accommodated at Tilbury. These include roll-on roll-off vessels, bulk carriers and container ships.

The rise of Tilbury has led to an increase in port industries in the area. The link to the M25 London Orbital Motorway means faster transport to other parts of the country. Now that the motorway around London is complete, it means that traffic no longer has to pass through the capital.

The building of this new port area has brought back life to the Thames. Without it, London as a port would have been finished. It has brought jobs and prosperity to this part of the Thames. It has led to an increase in other industries in the surrounding area. Again, this means more jobs.

Question

1 Using the information on the map in Figure 4.34, and in the book, describe why the older docks near London are declining. What can be done with docks such as these?

2 Describe why Tilbury was a suitable site for new docks.

3 Look at Figure 4.34. Name four different industries along the river. Why is it an advantage for some industries to be alongside river estuaries?

Airports

Flying is a fairly modern form of transport. In the 1950s, only a small number of aircraft were using a limited network of air routes. Planes were slower than the jets of today. Journeys tended to be bumpy because the aircraft could not climb above the turbulent cloud. Frequent landings were necessary for refuelling. All of this has changed as planes have become larger and more efficient. Passengers can watch films, have meals and travel in real comfort. Sometimes there is still some turbulence, but it does not last long. Modern planes can travel at great speed and go for long distances before having to refuel. There are now hundreds of air routes all over the world. Airlines carry millions of passengers and large quantities of freight each year.

Once airborne, planes have freedom to move around. But they still have to take off and land at airports.

Airports need flat sites for smooth take-offs and landings. The soil has to be well-drained, to prevent flooding. Airports take up large areas of land, so the sites chosen need to be cheap. In Britain, runways must be in an east–west direction so that planes take off into the wind. Airports also need to be located near to large towns or cities so that they are easily accessible to a large number of people.

The main airports in Britain are all close to the major cities. London has two major airports (Gatwick and Heathrow). Ringway Airport serves Manchester, while Prestwick caters for Glasgow. Figure 4.37 shows the location of the main airports.

Heathrow is the largest airport in Britain. It is situated 20 km from central London. Figure 4.45 shows its layout.

Gatwick is London's second airport. It is linked to London by the M23, as well as by a fast rail link every 15 minutes (the Gatwick Express). Nearby is Crawley New Town, which provides much of the workforce needed by the airport. Crawley also has numerous factories whose work is associated with the airport.

The number of people travelling by air is increasing all the time. So too is the amount of freight being transported. This means that more and more planes are using the airport. The increase is expected to continue. There were two possibilities open to the authorities in the early 1980s:
(a) to enlarge the existing facilities at Gatwick by building an extra runway and enlarging the present terminal buildings. People living near the airport are likely to object because more planes will mean more noise and danger;
(b) to build a new airport. Several sites were studied in detail, but it would cost an enormous amount of money to build a completely new airport. In some cases it would mean using up much farmland. That was one objection to building at Cublington. Foulness in Essex seemed to offer a better site as planes could take off and land over the sea. But this would have meant reclaiming land from the sea, as well as building

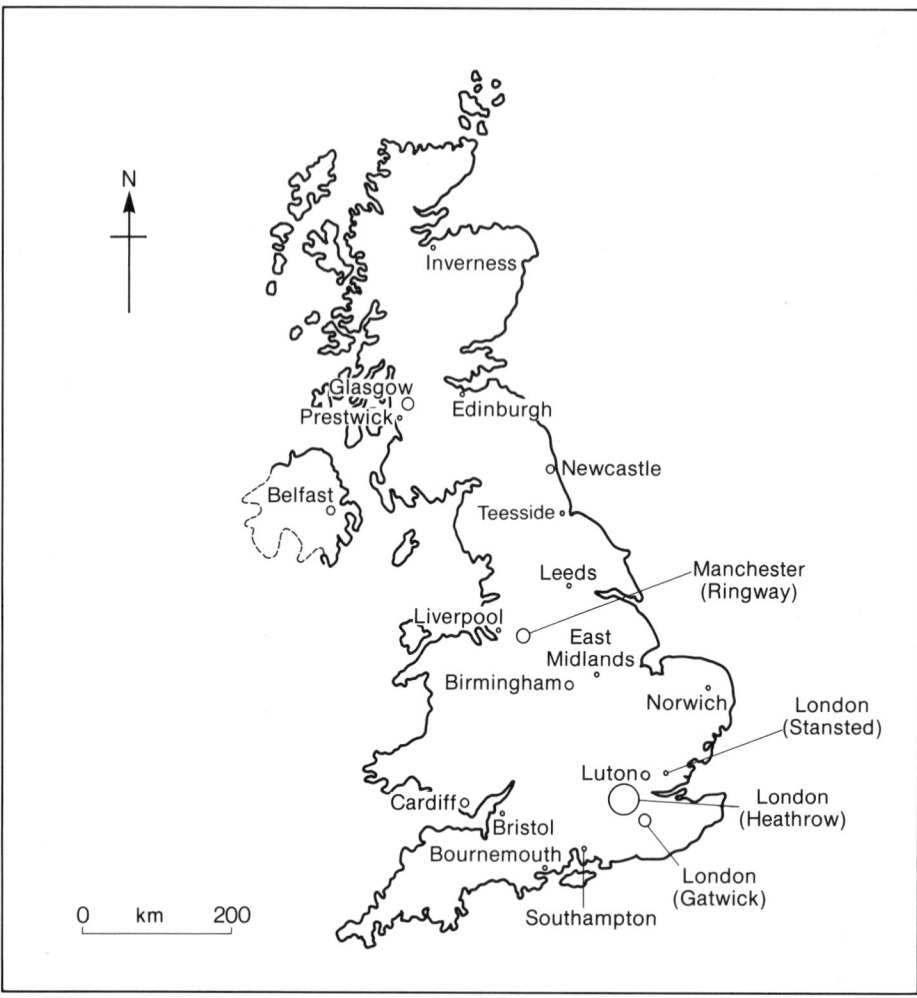

Figure 4.37. *The main airports in the United Kingdom*
The larger the circle, the greater is the number of passengers passing through the airport

Transport · AIRPORTS

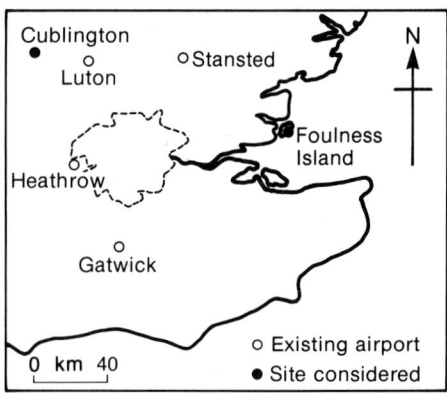

Figure 4.38. *London's airports, with sites considered for a third airport*

Figure 4.39. *Travel from London to Glasgow (1985)*

Type of transport	Time taken	Cost
Plane	1 hour 10 min	£57
Train	5 hours	£36.50
Coach	7 hours 10 min	£10.50

Figure 4.40. *Travel from London to Birmingham*

Type of transport	Time taken	Cost
Plane	45 mins	£41
Train	1 hour 36 min	£13.40
Coach	2 hours 15 min	£5.25

new road and rail links. A small existing airport, like Stansted in Essex, could be developed.

In 1985 the Government decided to make Stansted London's third airport. Stansted had been a small airport. It was near the M11 into London and the railway at nearby Bishops Stortford linked the airport to central London in 35 minutes.

The plan for Stansted is to have one main runway and two terminals, like Gatwick. By 1993 it is hoped that 8 million passengers will be able to use the airport. After that date, the number of passengers could rise to 15 million. Eventually, a total of 25 million might be expected to pass through each year. The airport will concentrate heavily on passengers although a small amount of cargo will be dealt with as well. About 90% of the passengers are expected to be tourists. (At Heathrow, about half of the passengers are travelling on business, the others being tourists.)

Some people still think that a third London airport will not be needed. They argue that planes of the future will be able to take off and land on much shorter runways. Some passenger-carrying planes may be able to take off vertically, like a helicopter. One proposal has been to turn one of London's docks into a short-runway airport called STOLPORT. (Short Take-Off and Landing).

Other people might support the idea of building a new airport, but do not want it built near them! People already living below flight paths have to put up with a great deal of noise and pollution, both day and night. The authorities have given some help towards the cost of double glazing and sound proofing.

Living near a busy airport is still not very pleasant.

Air travel is certainly the fastest form of transport over a long distance. For journeys of over 500 km it is much quicker than travel by either road or rail, as shown in Figure 4.39.

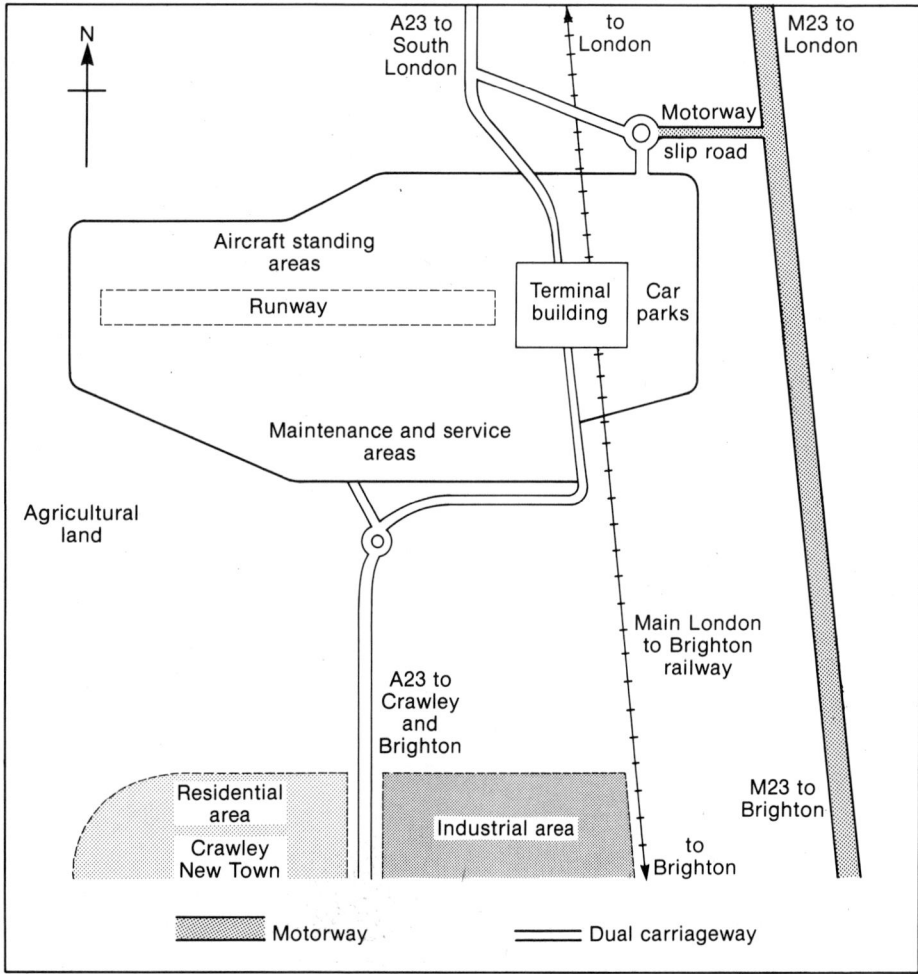

Figure 4.41. *Gatwick airport*

Over shorter distances, air travel may be slower than other forms of transport. This is because airports are located outside towns and cities. While road and rail users can travel from one city centre to another, air passengers have to travel to and from airports.

As well as the time shown in Figure 4.40, travel by air would also include 45 minutes of road or Underground travel to Heathrow, 15 minutes of rail travel from Birmingham airport to the city centre, plus the time taken to transfer on and off the plane. This means a total of at least two hours.

The amount of freight carried by planes has increased a great deal. They transport a wide range of goods. These include machinery, electrical goods and high value goods. Aircraft are also frequently used to carry 'highly perishable' fresh foods such as tomatoes, fruit and flowers.

Figure 4.43. *Gatwick airport*
The 'Satellite' or 'Pier Three' became fully operational in 1983. There is a fully automated elevated rail system between the Satellite lounge and main terminal. In 1986/87, Gatwick became the third largest international airport in the world.

○ HEATHROW AIRPORT

Year	Amount of cargo (tonnes)
1950	14 000
1960	93 000
1970	373 000
1980	531 000

Year	Number of passengers
1950	523 000
1960	5 381 000
1970	15 381 000
1980	27 771 000

Figure 4.42 *Heathrow Airport, cargo and passengers*

Figure 4.44. *Part of Heathrow airport*
Can you see any runways? airport buildings? road links?

Transport · AIRPORTS

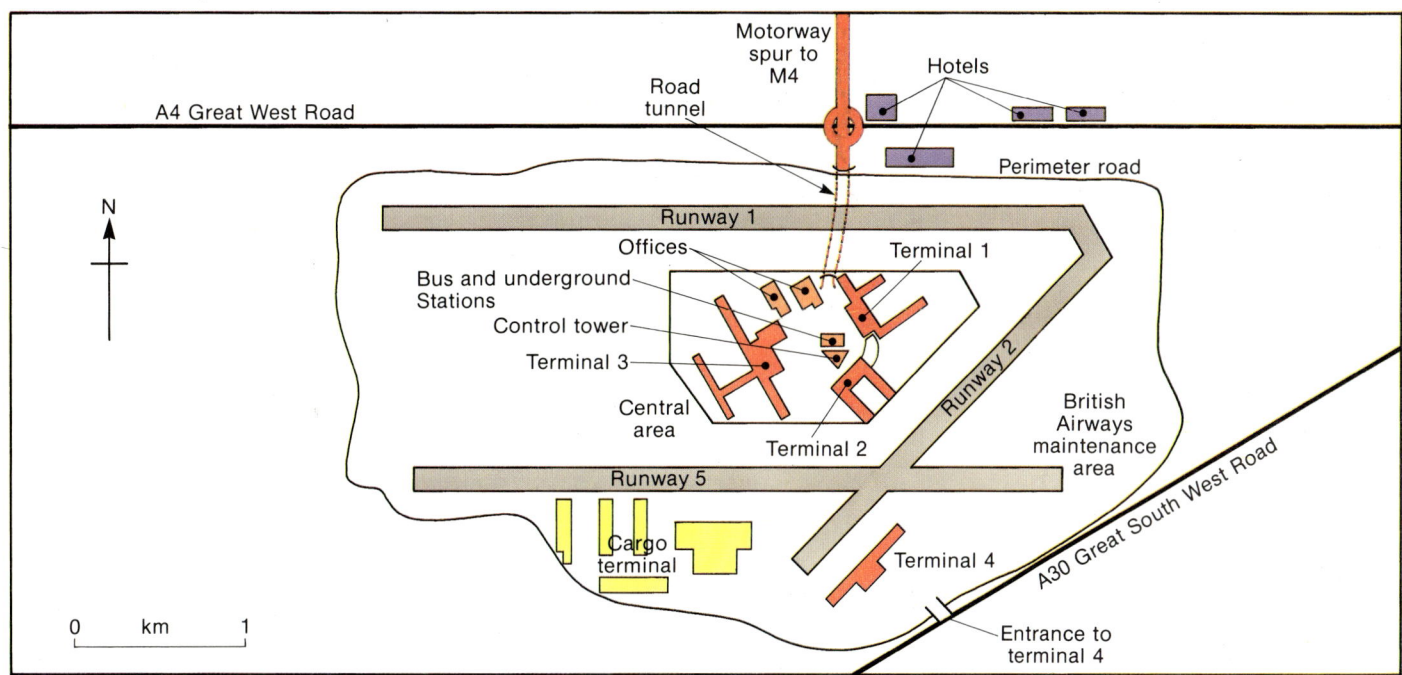

Figure 4.45. *Heathrow airport*

○ HEATHROW AIRPORT – SOME IMPORTANT FACTS

World's busiest international airport, with over 25 million international passengers every year.
Fourth largest airport in the world in terms of the total number of passengers (international and domestic).
Originally an aerodrome for experimental flying, then in 1944 used as a military airfield.
Opened for passenger flights and cargo planes in 1946.

It has four large terminals:
Terminal 1 is for local domestic services and short distance routes
Terminal 2 is for short distance routes mainly operated by foreign lines
Terminal 3 is for long distance routes
Terminal 4 a new terminal built in 1984

Heathrow can cope with a plane landing or taking off every minute.
Heathrow is the largest 'port' in terms of the value of cargo, in Britain.
There are fast links to central London by Underground or by express buses using the M4 motorway.
There are links with Gatwick airport by helicopter, by express coach and by rail.

Questions

1 Study Figures 4.41 and 4.43, then draw a labelled sketch diagram of the photograph. Clearly label the aircraft standing area, airport buildings and main roads.

2 Describe the types of industry connected with planes, that might be found in an area around an airport such as Crawley.

3 Heathrow Airport is the world's largest airport in terms of international traffic.

Using the information in Figure 4.42:
(a) State when the greatest increase in the number of passengers occurred.
(b) Describe and account for the trends shown in the table for both passenger and cargo traffic.

4 Millions of people use Heathrow each year:
(a) State the problems of large numbers of passengers travelling to and from the airport.
(b) Describe how the problems of transport are being overcome.

5 What conflicting interests had to be considered before it was decided to develop Stansted as London's third airport?

SECTION 5 · *Population and Settlement*

Earlier sections have shown how great changes have taken place in industry and transport during the past 100 years. The population has also altered. There are now many more people living in Britain and many more of them live in towns and cities. Such urban centres have each developed in different ways. Some are industrial, others are tourist centres or regional centres, while many are small market towns.

Population

In the year 1100, two million people lived in Britain. By 1700, the figure was 8½ million. By 1790, the population had risen to about 10 million. Yet by 1850, the total had jumped to 20 million. Thirty million was reached only 30 years later in 1880. Forty million people were living in the country by 1910. The figure did not rise to 50 million until 1957, so the rate of increase had slowed down. But today, with the population over 56 million, there are still nearly twice as many people living in Britain as there were a century ago.

○ THE CHANGING POPULATION

The rapid rise in population actually began in about 1800. Two things began to happen at the same time. There was a fall in infant mortality (deaths amongst babies) and a drop in the death rate. People began to live longer, on average.

We work out population figures using census returns. These are counts of the population held every ten years. They began in 1801.

By 1800, industry was growing in Britain. Many people were moving to towns in search of jobs in mills and factories. Housing conditions were usually poor but people generally had more money to spend on feeding and clothing their families. Towns grew because industry needed more and more workers. New houses and hospitals were built. Water supplies were improved and sewers were built. At the same time as doctors were finding new ways of treating diseases, the population's living conditions were getting better as well. As a result, fewer babies and children died young and adults started to live to an older age.

Nowadays, people are living even longer, as Figure 5.3 shows.

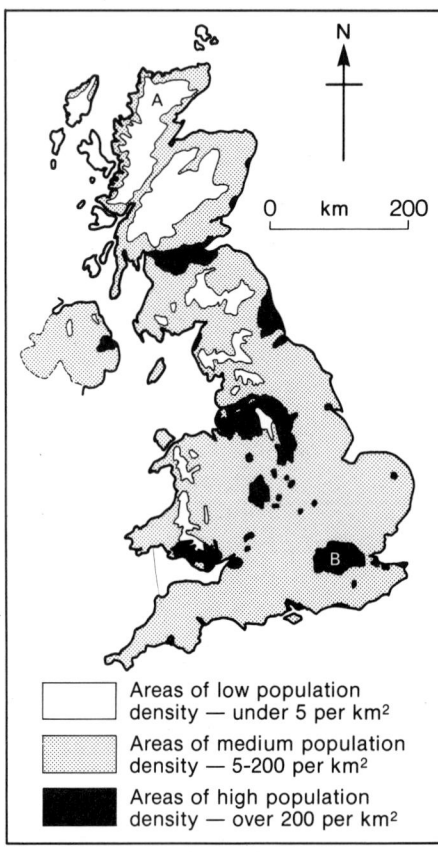

Figure 5.1. *Population density of Britain*

Figure 5.2. *UK birth and death rates*

	1851	1981
Birth rate	34	13
(births per 1000 of population)		
Death rate	22	12
(deaths per 1000 of population)		

Figure 5.3. *Average UK life expectancy*

	1851	1981
Male	40	69
Female	42	75

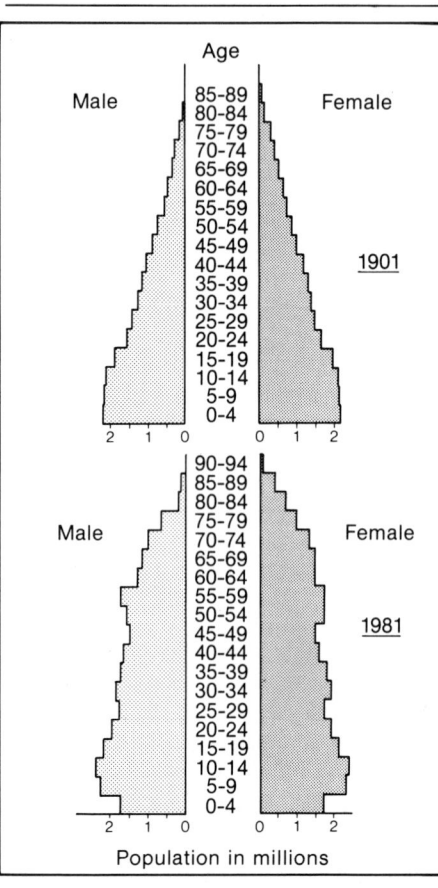

Figure 5.4. *Sex and age pyramids for the population of the British Isles*

MIGRATING POPULATION

Areas like the south-west and the south coast of England are especially popular retirement areas. They attract older people from all over the country. This sort of population movement also happens when people move to look for new jobs.

Over the last few decades, there has been a steady population drift from the north to the south. Industries like textiles, coal mining and iron and steel are in decline. These originally attracted workers to towns in Yorkshire, the north-east and south Wales. New industries, like light engineering, have been introduced in these areas to help provide jobs. But many people still prefer to seek work in the south.

In particular, the population of the south-east has increased considerably over the past 40 years. However, since the 1970s, more people have been preferring to live outside the London area. Many people are also choosing to live in the south-west. During the last 10 years, new industries have added greatly to this region's existing popularity as a retirement area.

This sort of population movement is called a migration. Such migrations have been particularly obvious in Britain since the 1950s.

Apart from the north–south movement described, there has been a continuing movement of people from the countryside to towns. Farm mechanisation is the main reason. There are fewer jobs available in farming areas. Another trend is people moving away from remote to more populated areas. For example, there has been a steady drift from the north to the south of Scotland. Such rural depopulation is a particular problem. Efforts are being made to attract new jobs to isolated areas, like the Highlands of Scotland.

Besides movement within Britain, there is also migration in and out of the country. In the 1980s, more people have been leaving Britain than coming in to live. The south-east of England is the area most affected by people arriving and leaving.

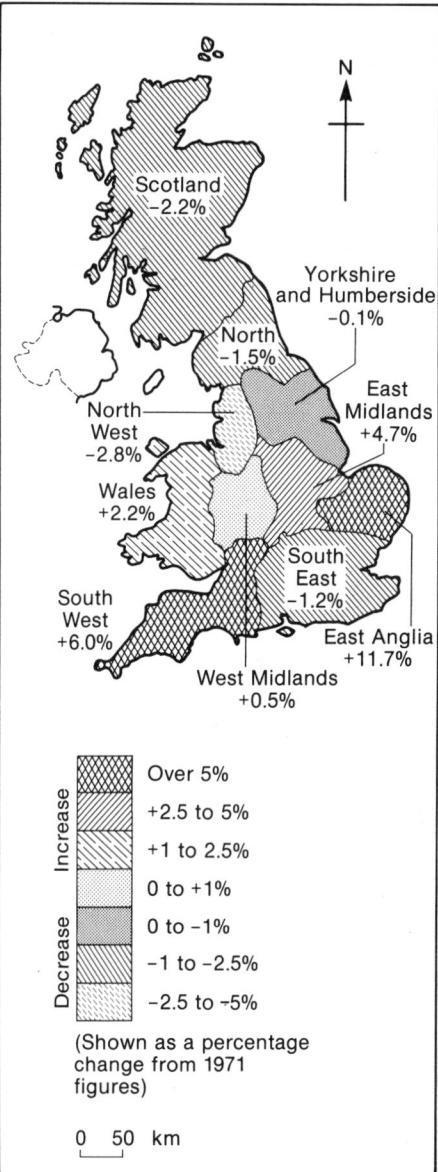

Figure 5.5. *Changes in population in Britain*
The map shows the percentage change between 1971 and 1981.

Questions

1 (a) From the information on page 96, draw a table to show how the population has increased during the last 900 years. The left hand column should be headed 'Year' and the right hand column headed 'Population'.
  (b) Draw a graph to illustrate the rise in population in Britain since 1100.

2 With reference to Figure 5.1, state the reasons why:
  (a) area A has a low density of population;
  (b) area B has a high density of population.

3 Study the age and sex pyramids of Britain in 1901 and 1981. What are the main differences between 1901 and 1981, as shown in the pyramids?

4 How does a higher proportion of older people affect demands on health and other social services?

5 Study Figure 5.5. State the area with (a) the highest decline of population and (b) the highest increase in population. Describe the distribution of areas showing an increase in population and those showing a decrease in population. (Note that the 1981 census showed the first ever decline in the population of south-east England).

6 What are the main areas of higher unemployment? (Look again at Figure 3.51.) Suggest reasons why some areas have a higher unemployment rate. What extra facilities are needed to cater for the unemployed?

Settlement

A settlement is where people live. It may be an isolated farmstead or a small group of houses in the country. These are called rural settlements. Most of us live amongst larger groups of people in towns or cities. These are called urban settlements. They may have thousands, hundreds of thousands or millions of inhabitants.

Wherever they live, people need certain services. These include means of transport, food, clothing and items bought less frequently, like furniture and electrical goods. They also require social services which include libraries, churches and doctors' surgeries.

In rural areas, villages can provide many basic daily requirements like bread and groceries. Village shops act as a centre for many farms and hamlets. But people wanting do-it-yourself equipment, hardware and clothing must usually be prepared to travel further afield, to the nearest town.

It may be necessary to visit an even larger centre, like a city, for major items like expensive clothing, furniture or hi-fi equipment. Small towns may be able to offer a limited range of such goods. Only the largest towns may be expected to provide a comprehensive selection of more luxury items.

In large urban areas, small shopping parades provide many basic local needs. Twenty or so shops in a small neighbourhood shopping centre can offer a wider range of items. The town or city centre with several hundred shops, including department stores, is able to provide the most complete choice of goods and services.

Figure 5.6. *The village of Llangwn in Clwyd, Wales. There are about 20 houses, one shop/post office, a chapel, a junior school and no pub. What reasons would villagers have for going to a larger settlement?*

Figure 5.7. *Part of a city centre*
This is a pedestrian precinct. Explain why it is a benefit to shoppers to have such precincts.

○ ORDER OF SETTLEMENTS

Settlements range in size from the smallest farmstead, to the capital city with several million people living there. The following is a progressive list, starting with the smallest leading to the largest. It is known as the hierarchy of settlements.

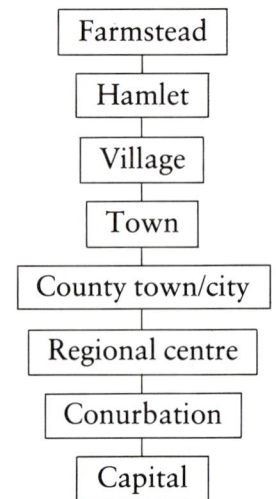

Farmstead
Hamlet
Village
Town
County town/city
Regional centre
Conurbation
Capital

Although it might seem easy to distinguish between the different types of settlement, in practice it is often quite difficult.

Population and Settlement · SETTLEMENT

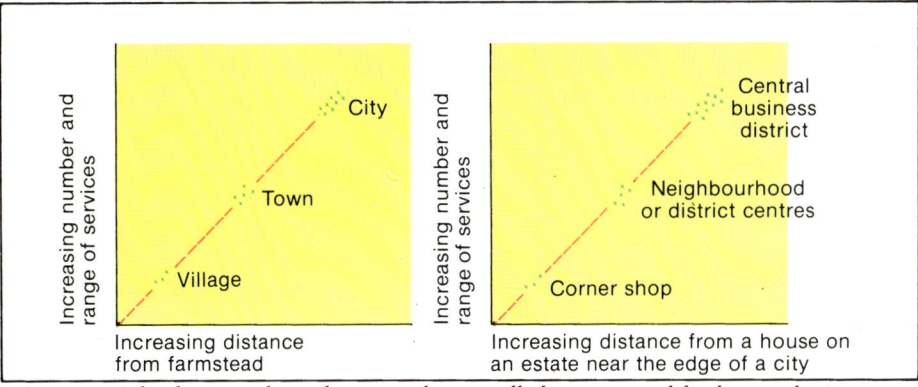

Figure 5.8. *The diagram shows how people generally have to travel further to obtain a greater range and choice of goods and services. The first diagram shows how a farmer will have to travel further to find a greater range of goods. Describe what the second diagram shows.*

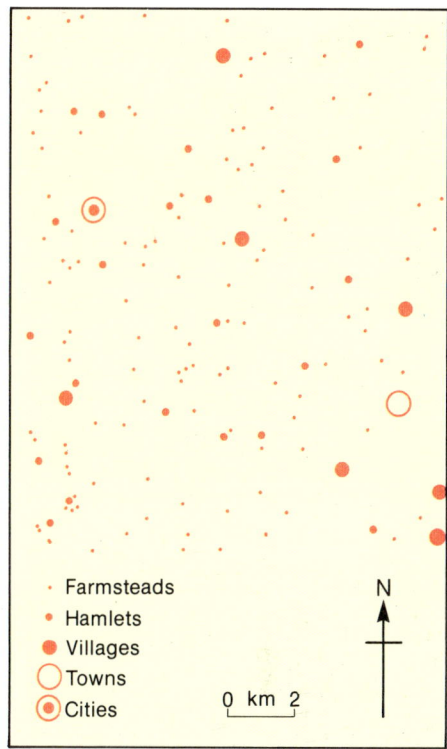

- Farmsteads
- Hamlets
- Villages
- Towns
- Cities

Figure 5.11. *Distribution of settlement Describe the pattern shown on the map. Include in your account the approximate numbers of each type of settlement, and how they are arranged.*

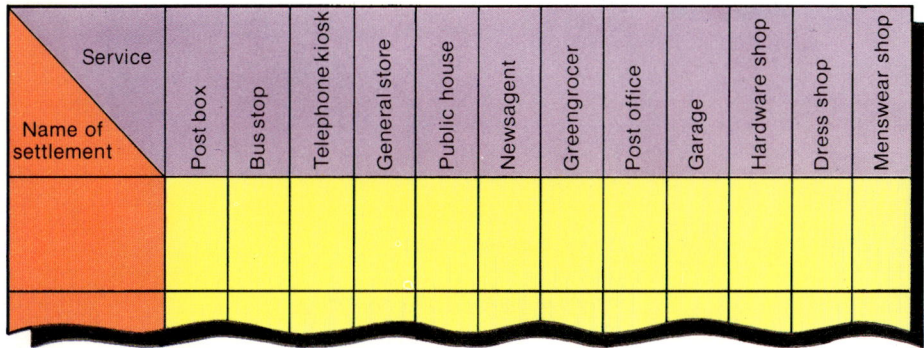

Figure 5.9. *Settlements and their services*

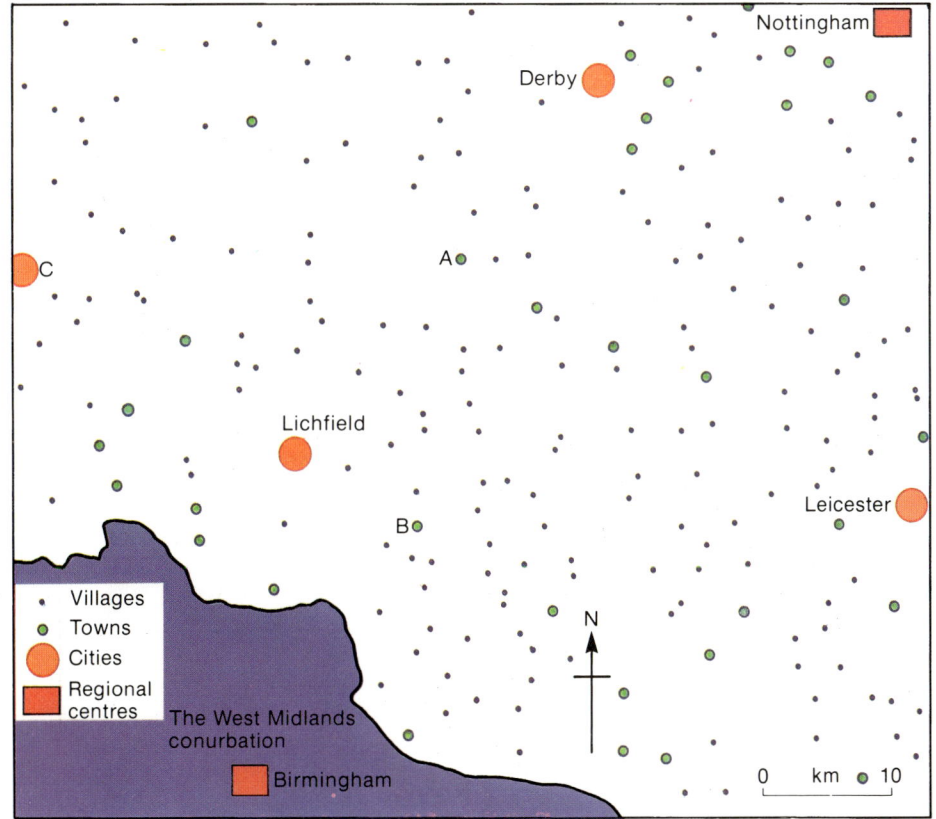

Figure 5.10. *Patterns of settlement in the North Midlands Using your atlas, find out the names of towns A and B, and city C. Describe the distribution of settlements shown on the map.*

Questions

1 Describe for yourselves what you understand by the settlement terms: hamlet, village, town. Include in your answer the services you would expect to find in each, and the population size.

2 Think of examples of settlements that you know yourselves. Copy out the table in Figure 5.9. Now tick any boxes where each particular service exists for the settlement you have chosen. For instance, one settlement's services might only consist of a pillar box, bus stop and telephone kiosk, so only three boxes will be ticked.

Now, by looking at the table you should be able to draw a line to distinguish between these settlements. How does it differ from your original decisions?

THE NORTH MIDLANDS

The two diagrams in Figures 5.10 and 5.11 show the patterns of settlement in part of the North Midlands. Figure 5.11 represents a small area of land. Its boundary is marked on Figure 5.10 by a dashed line. There are a large number of lower order settlements, like hamlets and villages, but only one conurbation. Lower order settlements are small and have only very few services. Higher order settlements are the opposite, usually having a high population and a large number of services.

In the past, farmsteads and hamlets tended to rely on villages for some services. The villages depended upon the towns and cities for many of their goods and services. All looked to the regional centre to provide a wide range of facilities and amenities. Thus the larger settlements were always surrounded by those lower in the order. To some extent this is still true, but modern transport has made higher order centres much more accessible than previously.

The ideal pattern is where a settlement of a higher order lies in the centre of a hexagon, with settlements of a lower order at each corner. This should mean that the settlement at the centre is accessible to as many people as possible (see Figure 5.12).

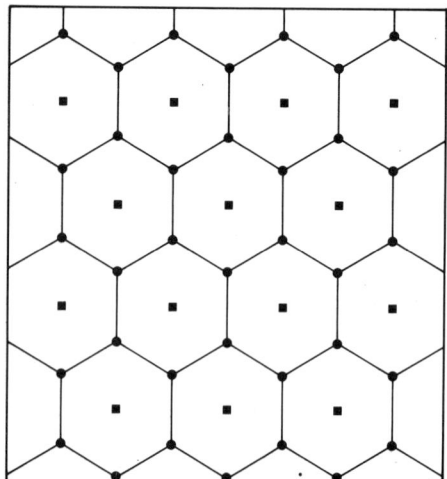

Figure 5.12. *Hexagon diagram*
The higher order settlements (squares) are surrounded by six lower order settlements (circles).

Returning to the maps, there appears to be a pattern of settlements, although the hexagons are nowhere near perfect. On the maps, although it is not very clear, you should be able to see that the larger settlements are fairly well spaced apart. There are more towns than cities, and the even larger number of villages are also well spaced apart.

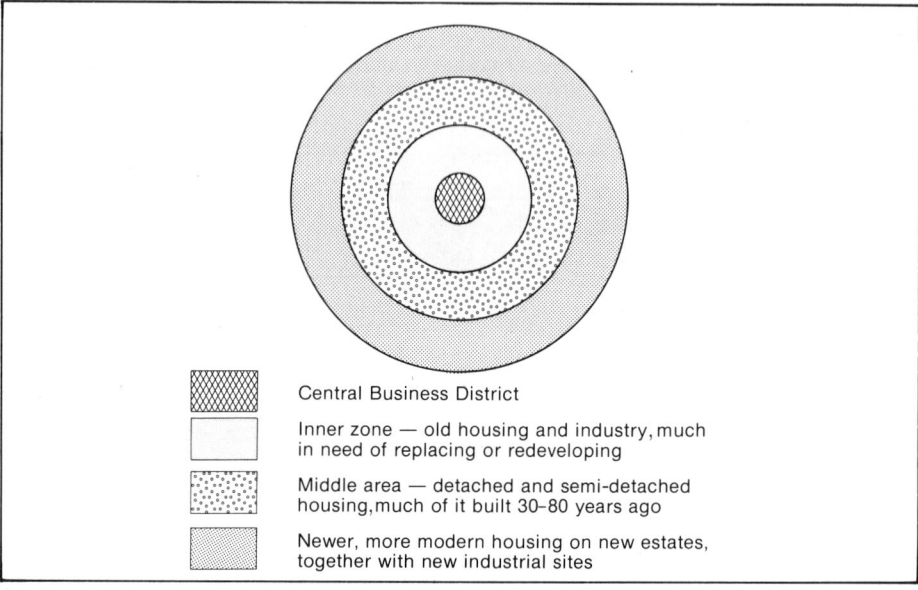

Figure 5.13. *Zones of a typical town*

Exercise

Copy out the table below. Use an atlas, and add your own examples on the right hand side. Choose different examples to those mentioned in this book.

Function	Example
Capital	
County town	
University town	
Cross Channel port	
A commercial port	
Market town	
Cathedral town	
Regional centre	
A railway town	

FUNCTIONS OF TOWNS

All towns provide services, but some towns have developed special functions. Some are market towns, where people from rural areas come to shop. Others have grown up around a university or a cathedral. A number of towns around the coast have become ports or holiday resorts. A few have developed into

Map reading questions

1 Look at Map 1 on page 123. What services do the following settlements have?
(a) Thornton-le-Moors (4474)
(b) Stoak (4273)
(c) Little Stanney (4174)
(d) Elton (4575)
How would you classify these settlements?

2 Look at Map 3 on page 125. What services do the following settlements provide?
(a) Brereton (0516)
(b) Mavesyn Ridware (0816/0817)
(c) Longdon (0814)
Which do you think they are: hamlet, village or town? Give reasons for your choice.

county towns with administrative offices. Others, like Oxford which is both a university and a county town, combine several functions. Figure 4.4 on page 79 shows Oxford to be a route centre as well. New Towns have a function. They have been set up by the Government to provide homes and work so that people can move out of crowded cities.

○ INSIDE TOWNS

Figure 5.15 shows how a typical town might be arranged. There are a series of rings or zones which have developed over many years.

Central Business District

The innermost area is the Central Business District (CBD). This is the main commercial area and contains a wide variety of buildings. There are many shops, ranging from small retailers selling items like shoes and clothing, to big department stores. A large number of banks, building societies, estate agents and insurance brokers have businesses in this zone.

Businesses in the CBD have to pay very high rents and rates. The amount is based on the area of floor space and the frontage of the building. Banks, building societies and large stores can afford high costs. Consequently, they usually seek the best high street sites in towns. These are often corner sites, where the pedestrian routes meet. Many small shopkeepers are unable to afford the high costs of running a business in the CBD. A number have been forced to seek cheaper sites outside the centre.

As land is expensive and new development is limited, single storey buildings are unusual. There is storage space or office accommodation over most shops. Very high office blocks have been built in many CBDs.

Due to the increasing number of people travelling into the CBD for shopping, large multi-storey car parks have been built in the town centre. This has helped to reduce parking and traffic congestion in busy side streets.

Inner Zone

The next area surrounding the CBD is the inner zone of old housing and industry. Much of this was built during the last century. Many of these rows of terraced houses are in a very poor state of repair. But they provide high density homes for many people who cannot, or do not, wish to move.

Middle Zone

The third area is usually made up of housing built during the first half of this century. Builders tended to construct long rows of detached and semi-detached houses, mostly with front and rear gardens. Although such housing was probably of a better standard than that in the inner zone, there was little planning control over what was built. Later, the planning authorities encouraged the use of crescents and long winding roads. Much valuable farmland was used up by the great increase in housing. This was known as 'urban sprawl'. An extension in suburban railways and bus routes also meant people could travel further to their places of work.

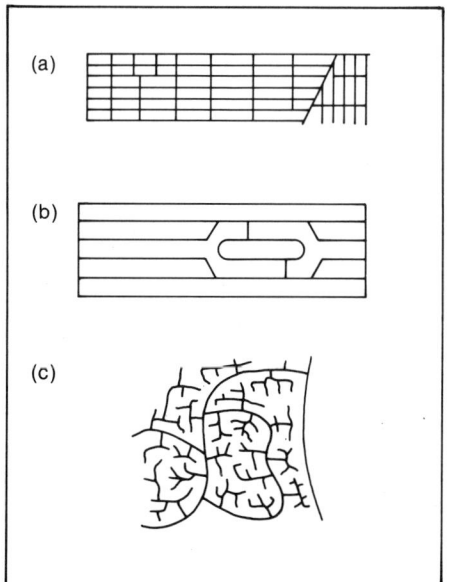

Figure 5.14. *Road patterns*
(a) Grid iron pattern
(b) Geometrical pattern
(c) Infilling pattern

Outer Zone

The final stage took place from the late 1950s. Carefully planned estates were built. Long rows of houses became unpopular. Builders were encouraged to build small groups of dwellings around cul-de-sacs. New estates are now likely to have their own parades of shops. The main roads through such developments are often curved. This is intended to reduce the speed of traffic through restricted areas. Open spaces are provided where people can walk and children can play.

The three diagrams in Figure 5.14 show the road patterns associated with the last three zones of housing:

Grid iron pattern   Back-to-back terraced housing in blocks. Much of this housing is pre-1900 and is characteristic of the inner zones of towns and cities.

Geometrical patterns   Houses spaced further apart with gardens and long straight roads and later, circular and oval crescents.

The infilling pattern   This was adopted after 1950. Several important linking roads with long sweeping corners (called artery roads) take traffic through the estate. Filling in the spaces are a large number of cul-de-sacs. Each has 12–24 houses. This occurs in modern neighbourhood units of New Towns.

Questions

1 Look at your own town, or the town nearest to you. Are there zones which look different and have different functions? Describe each zone. How do the zones differ from each other? Can you find any of the road patterns shown in Figure 5.14?

2 London has special functions as a capital city. What buildings in London tell you that London is a capital city?

Conurbations

In some parts of Britain, towns have developed in groups. Some are as little as 10 km apart. The cotton towns to the north of Manchester are a good example. Originally such towns were small, and separated from each other by green fields. As they have grown larger, the fields have disappeared under new housing and industries. Now one town often runs into the next. However, each usually retains its own Central Business District, a separate shopping area and its own character.

A conurbation can thus be described as a group of towns that have grown into each other to form a continuous built-up area.

A conurbation has a population of over 1 million people. Examples include 1.5 million living in Merseyside and 2.5 million living in the West Midlands. But the number of people living in the major conurbations has been declining now for several years.

Problems of living in conurbations
Poor housing conditions are an important reason for this drop in the population of conurbations. There are still large numbers of old back-to-back terraced houses in most inner city areas. Some new housing has been built within the conurbations. But many people have had to move away from the inner areas. Some have had to leave the conurbation altogether in search of a new home. Even more people will probably have to do so in the future.

As many more people now need to travel back to the conurbations to work, a more efficient transport system is needed. Roads in particular need to be improved if they are to cope with the extra traffic. New Metropolitan Councils were formed in the early 1970s to coordinate future planning for both transport and housing in conurbation areas. New road schemes, like urban motorways, dual carriageways and ring roads have been introduced in some areas. A Passenger Transport Executive has been appointed in each conurbation. The Metropolitan Authorities were abolished in 1986.

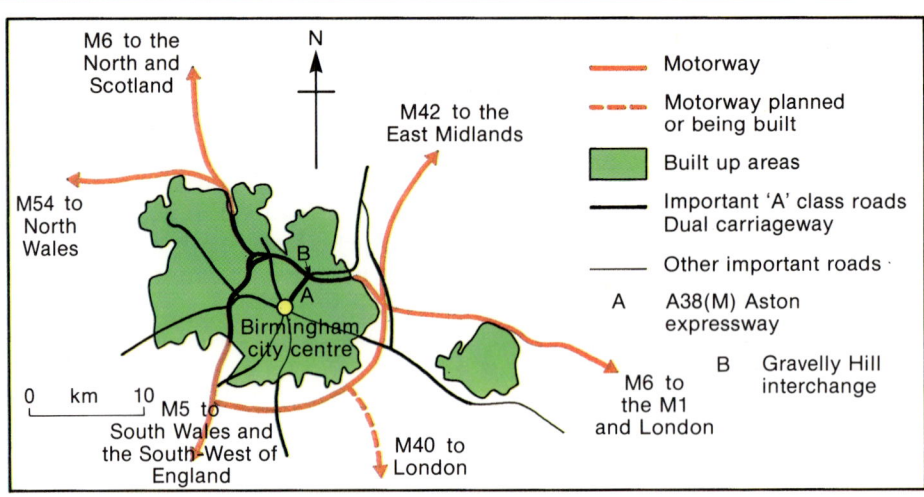

Figure 5.15. *The motorways around Birmingham*

○ WEST MIDLANDS CONURBATION

One of the biggest conurbations can be found in the centre of Britain. The West Midlands is situated around Birmingham. It stretches from Wolverhampton in

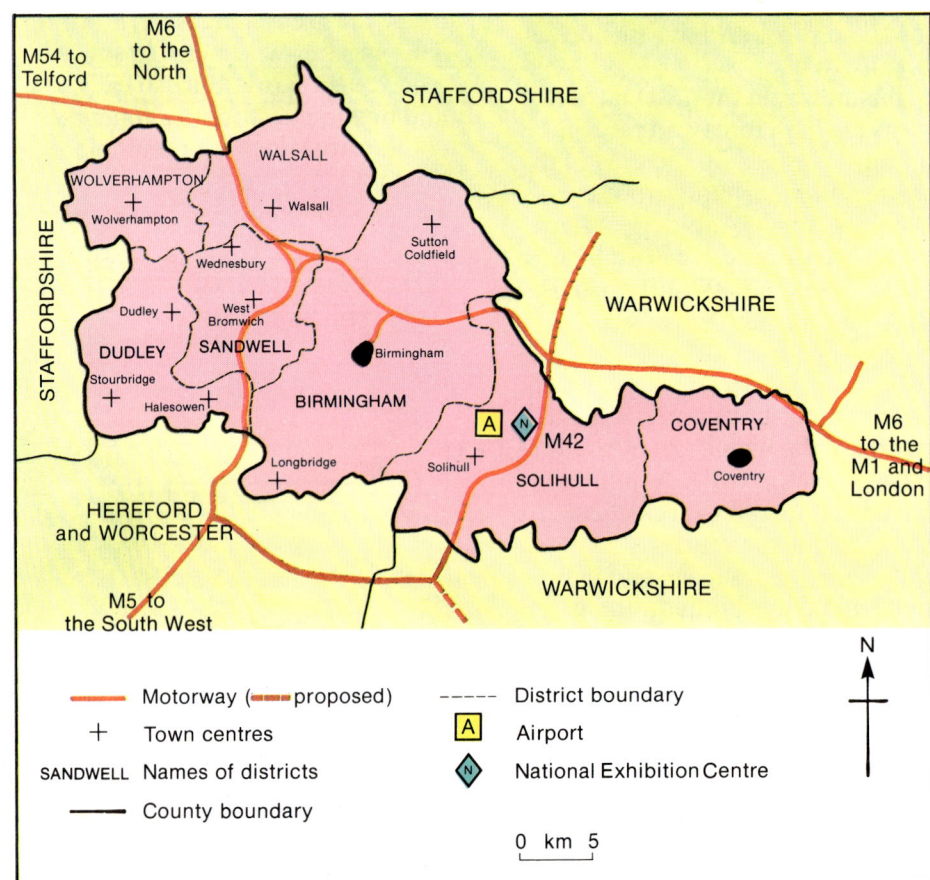

Figure 5.16. *West Midlands conurbation*

the north to Coventry in the south. It is a heavily built up area, linking housing, services and industry. Much of the industry involves manufacturing. Some industries have a long tradition, like jewellery making in Birmingham. The new industries are very varied. There is an emphasis on manufacturing parts for the car industry. But there are numerous other industries, particularly connected with metal working, for which this area is famous.

Figure 5.16 shows how the area is divided up. The main towns are marked. These were once separate, but they have all expanded into each other. All those in the north are traditional industrial towns.

Each town has its own shopping centre offering a good range of facilities. Several towns like Wolverhampton and Walsall offer a complete range of shops including department stores. The main centre, Birmingham, has all the department stores that normally would be found in a regional capital. Each town has its own cultural centres, recreational areas and administrative offices.

Throughout the West Midlands, great emphasis has been placed on creating an efficient road system (see Figure 5.15). The West Midlands is the junction of two great motorways: the M6 and the M5. The M6 links the north to the M1, and the M5 stretches from London to south Wales and the south-west.

Linking the motorways to other areas of the West Midlands are a number of fast roads, often dual carriageways. Some lead directly to Birmingham city centre. The most famous of these is the Aston Expressway. This takes traffic from Gravelly Hill interchange (Spaghetti Junction) to the city centre.

Figure 5.17. *The city centre of Birmingham*
The central business district is shown. The inner city relief road or 'ringway' completely encircles the district.

104 *The British Isles: Themes and Case Studies*

Over recent years there has been a decline in the number of people living in the West Midlands conurbation. This is true of all conurbations. Figure 5.18 shows this decline. As old houses are cleared for redevelopment, many people are encouraged to move to overspill areas outside the conurbation boundaries. Some people go, but some remain. Both groups often miss their neighbours and relatives. Old housing areas lack many amenities but they are often very friendly places.

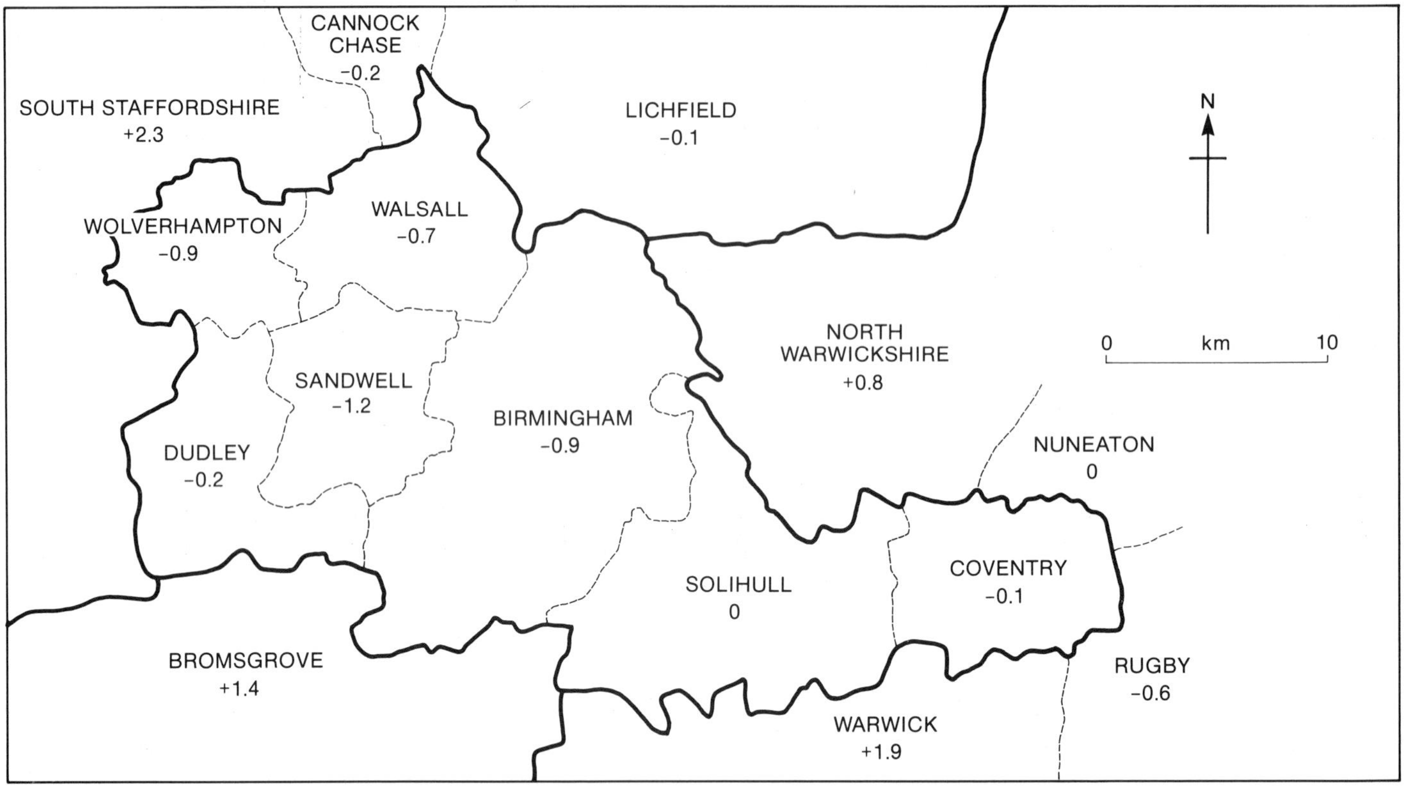

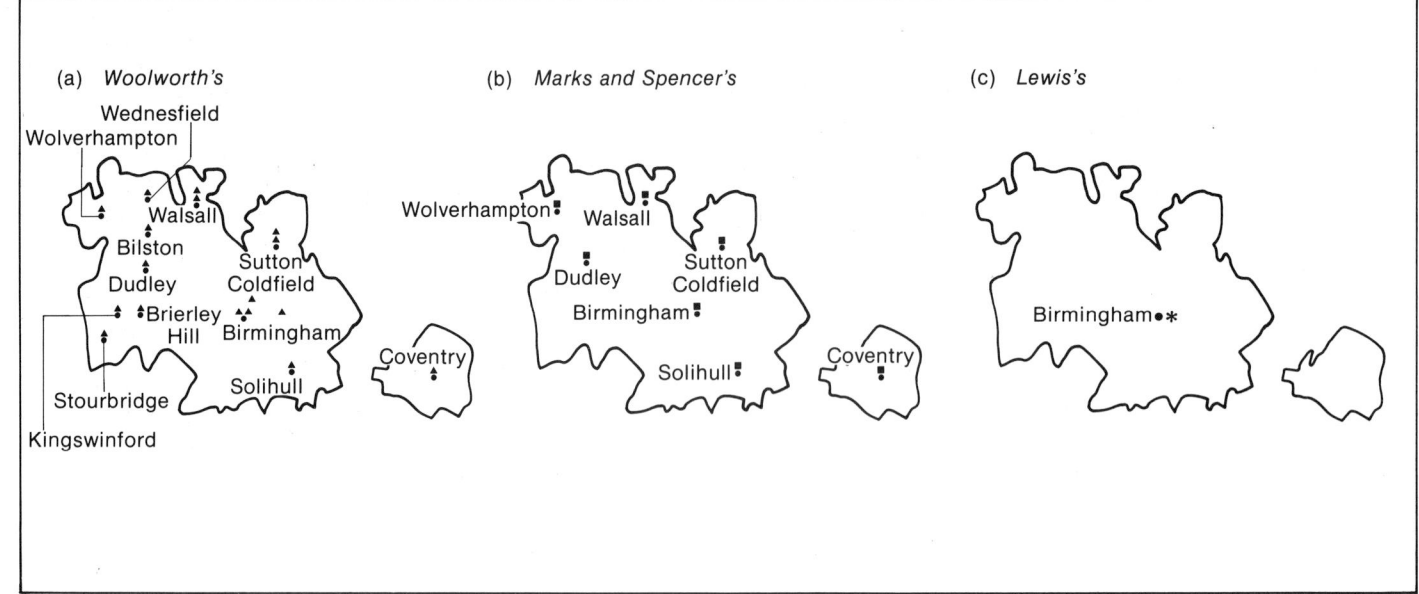

Figure 5.18. *Average percentage changes in population of the West Midlands per year, 1971–1984*

Figure 5.19. *Distribution of certain department stores in the West Midlands*
  (a) *Woolworth's*
  (b) *Marks and Spencer*
  (c) *Lewis's*
(a) *What do you notice about the numbers of each type of store shown in the three diagrams?*
(b) *Describe the areas that each type of store serves.*
(c) *Would you suggest that there is a similar pattern in conurbations that you know?*

Population and Settlement · CONURBATIONS

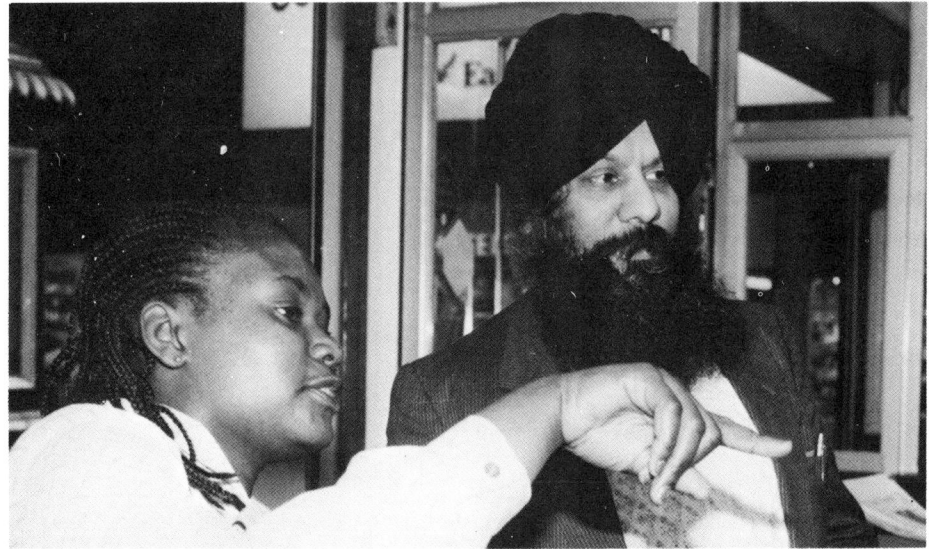

People miss the feeling of a community when they move to modern housing estates. For those left behind, life becomes very different as houses around them are demolished or boarded up.

The West Midlands Metropolitan Council was set up to coordinate all slum clearance and new housing schemes. The council also planned for an improved transport system and helped industry.

Some conurbations in Britain are increasingly multi-ethnic, with people from different cultures. This adds to the rich variety of community life in Britain. Areas must adapt to changing needs as populations change.

	Sutherland St. Area (left)	Gardens Rd. Area (right)
Population density per hectare	94	72
Percentage of population in shared accommodation	8.9	1.7
Percentage of population without or sharing a bath	9.3	1.0

Figure 5.20. *Old and new housing. The photo shows part of an inner-city area. Some old housing has been demolished and new dwellings have been built (right of photo). Describe the types of buildings found around Sutherland Street (on the left) and Gardens Road (on the right). Describe the areas of open space around Gardens Road. What disadvantages are there to the lack of open space around Sutherland Street? What disadvantages are there to the types of buildings around Gardens Road? In many inner-city areas, authorities are trying to renovate rather than demolish more older houses. Suggest reasons for this.*

Questions

1 Suggest reasons why large areas of conurbations are losing population.

2 Figure 5.18 shows the changes in population in the West Midlands as a percentage each year. The continuous black line marks the boundary of the former Metropolitan Council.
(a) Which area has the greatest decline in population?
(b) How many areas have a steady population or are increasing in population?
(c) Which area has the greatest increase outside the former Metropolitan Council?
(d) Write a couple of sentences to explain what is happening outside the former Metropolitan boundary, compared with inside the boundary?
(e) Name the county authorities that surround the West Midlands.

3 Study an atlas map of Britain, then decide what areas of the country could be referred to as conurbations. Make a list of these centres.

New Towns

In the late 1940s, there was a heavy demand for housing. Many inner city areas had been badly bombed during the Second World War (1939–1945). Other houses were very old and needed replacing. The Government decided to try and solve the housing problem by building some 'New Towns'.

The first group of New Towns was built around London. It was hoped that they would rehouse many of the people who had once lived in the capital. Other large cities had similar housing problems. New Towns were planned in other parts of the country.

Figure 5.22 shows the New Towns that were built around London. It was decided that they should

(a) be 35–50 km from the centre of London;
(b) have populations of between 50 000 and 80 000;
(c) be built on the outer edges of the Green Belt (land which had not been built on before);
(d) have their own shopping areas, industry, parks and open spaces.

Figure 5.23 shows the position of similar New Towns in other parts of Britain. Draw a table and list the New Towns that are within 50 km of (a) London, (b) Liverpool, (c) Birmingham and (d) Glasgow.

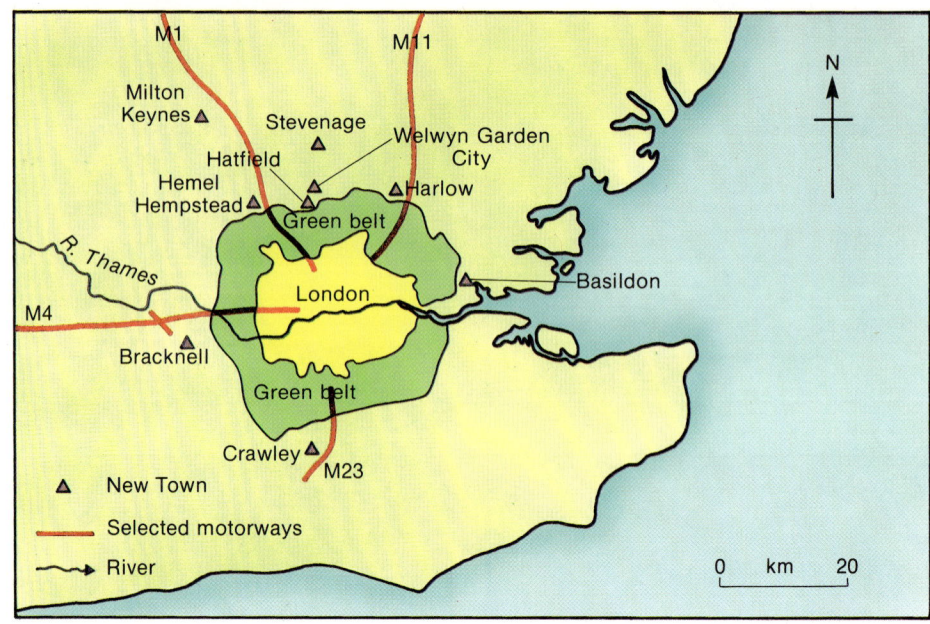

Figure 5.22. *New towns around London*

Layout of New Towns

Many old towns had lots of small, terraced houses. These had little or no garden and were built next to dirty factories. In contrast, the New Towns were carefully planned. Industry was sited well away from housing areas, with plenty of open spaces for recreation.

The plan of Crawley shows how the early New Towns were laid out (see Figure 5.24). There are twelve large areas of housing, called neighbourhood areas, a town centre and an industrial area near Gatwick Airport.

Figure 5.21. *Neighbourhood housing in Crawley*

Figure 5.23. *New towns of Britain*

Population and Settlement · NEW TOWNS

Each neighbourhood unit has its own small shopping centre for everyday needs. This can be seen on Figure 5.25, which is a plan of one neighbourhood area in Crawley – Furnace Green. The main roads leading through Furnace Green are called Weald Drive and Hawarth Avenue. These are known as artery roads. They are curved to stop the traffic going too fast.

New Town planners also thought about pedestrians. They often planned separate routes for walkers. Several towns have a network of footpaths linking the town centre with the housing areas. In Runcorn there is also a separate road for buses.

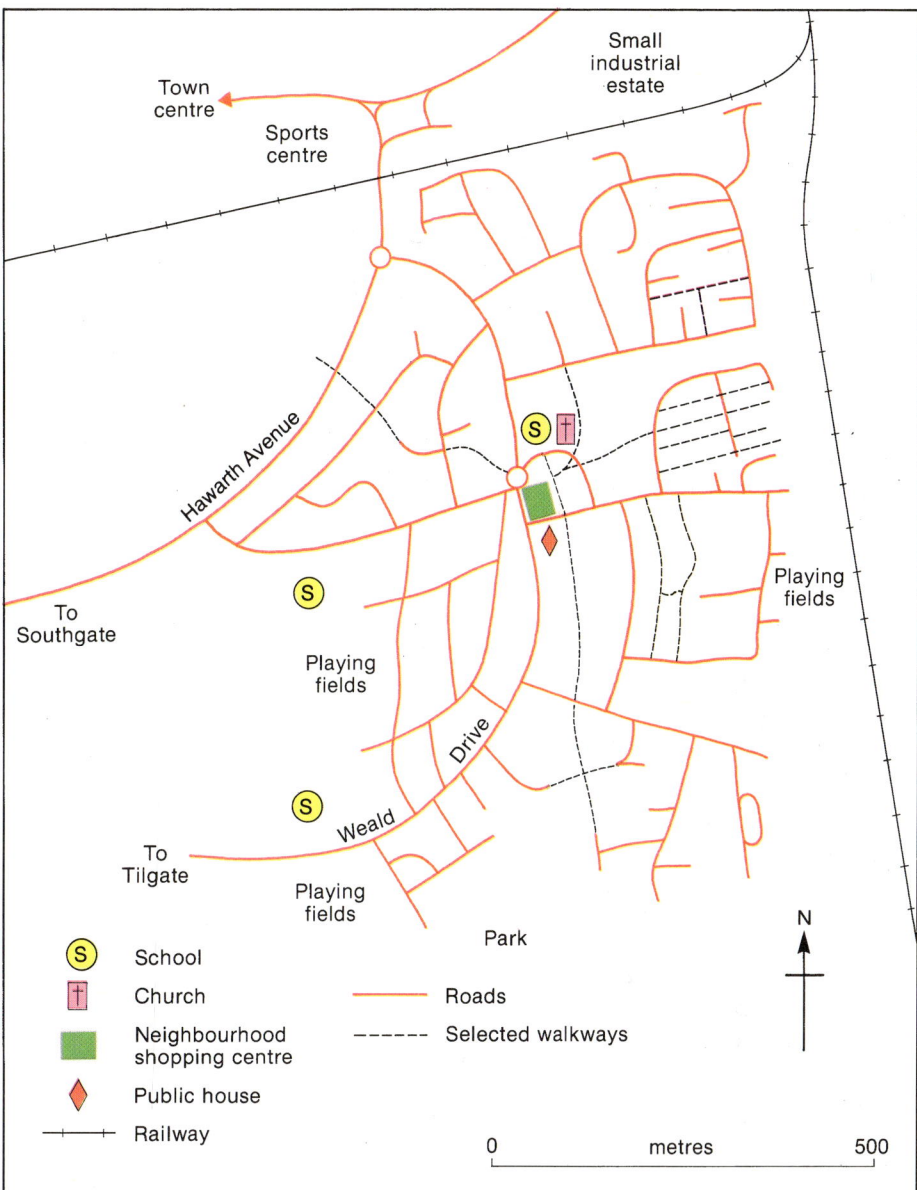

Figure 5.25. *Furnace Green neighbourhood area*

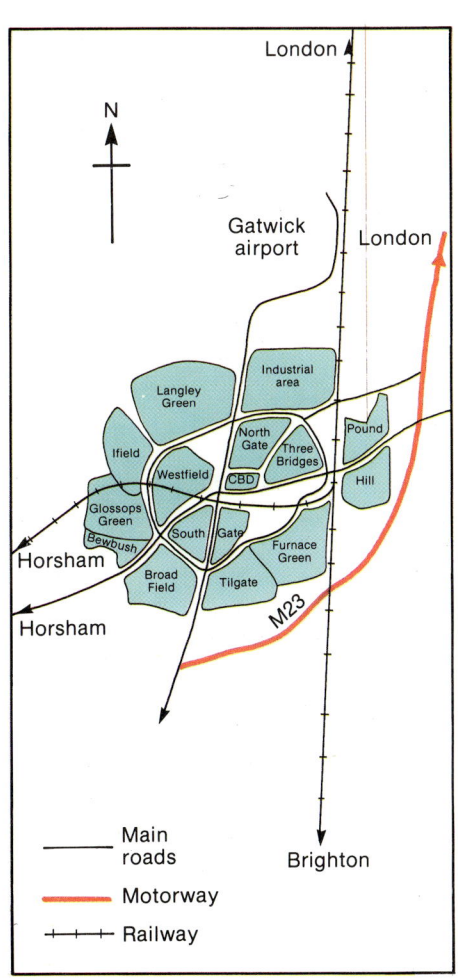

Figure 5.24. *Plan of Crawley showing neighbourhood areas*

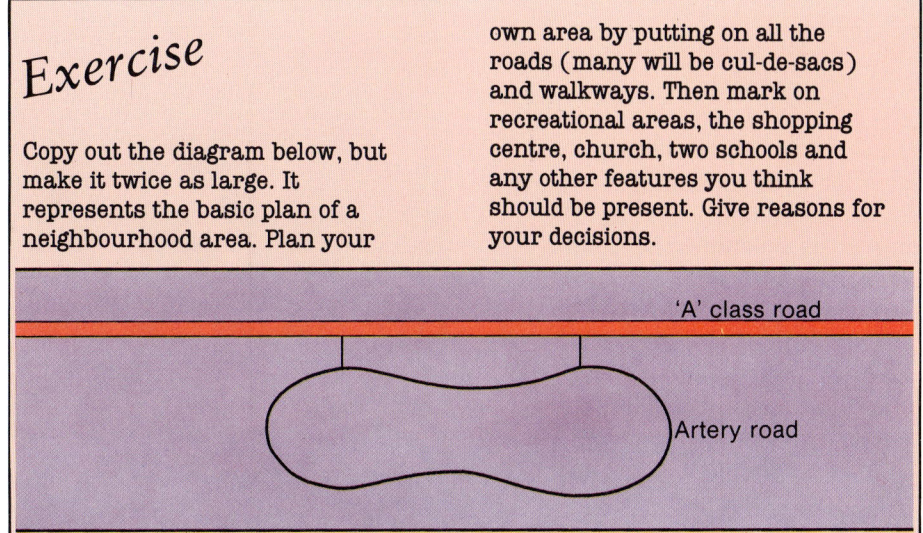

Exercise

Copy out the diagram below, but make it twice as large. It represents the basic plan of a neighbourhood area. Plan your own area by putting on all the roads (many will be cul-de-sacs) and walkways. Then mark on recreational areas, the shopping centre, church, two schools and any other features you think should be present. Give reasons for your decisions.

○ SKELMERSDALE

Skelmersdale New Town lies 25 km from Liverpool. It was built near the site of an old coal mine. The area was very rural. It was therefore referred to as a 'green field' situation. Many people came from Liverpool and the nearby town of Wigan, where there were many old houses that needed pulling down. Skelmersdale offered new houses, good shopping facilities and new industry.

The planners included the following points when designing the town's layout:
(a) There would be a high level of car ownership, requiring a good road network.
(b) A system of footpaths leading into the town should be provided, so that people could walk from any area to the main shops without crossing roads.
(c) The New Town should be compact, so that all housing can be within 1½ km of the town centre. This would mean less need for a neighbourhood system.
(d) There should be several industrial areas within easy reach, and little traffic congestion when the factories opened and closed.
(e) There should be a balanced population, with people of all ages.

The New Town was planned in 1961, and building began in 1964. It is hoped that the population will reach over 40 000.

The layout of Skelmersdale

The plan of Skelmersdale in Figure 5.26 shows its basic layout. The central area includes the town centre, bus station, main churches, police station, library and sixth form college.

The main housing areas are linked to each other, and the town centre, by roads and separate walkways. There are some dual carriageways for fast access, smaller estate roads and cul-de-sacs. No houses lead directly on to the fast access roads. The less important roads are deliberately curved to slow down traffic.

Industry in Skelmersdale is very varied. It is mostly light industry, and includes the manufacture of metal cans, tools, heat insulation materials, furniture, cosmetics and clothing. The factories are new and many are purpose-built.

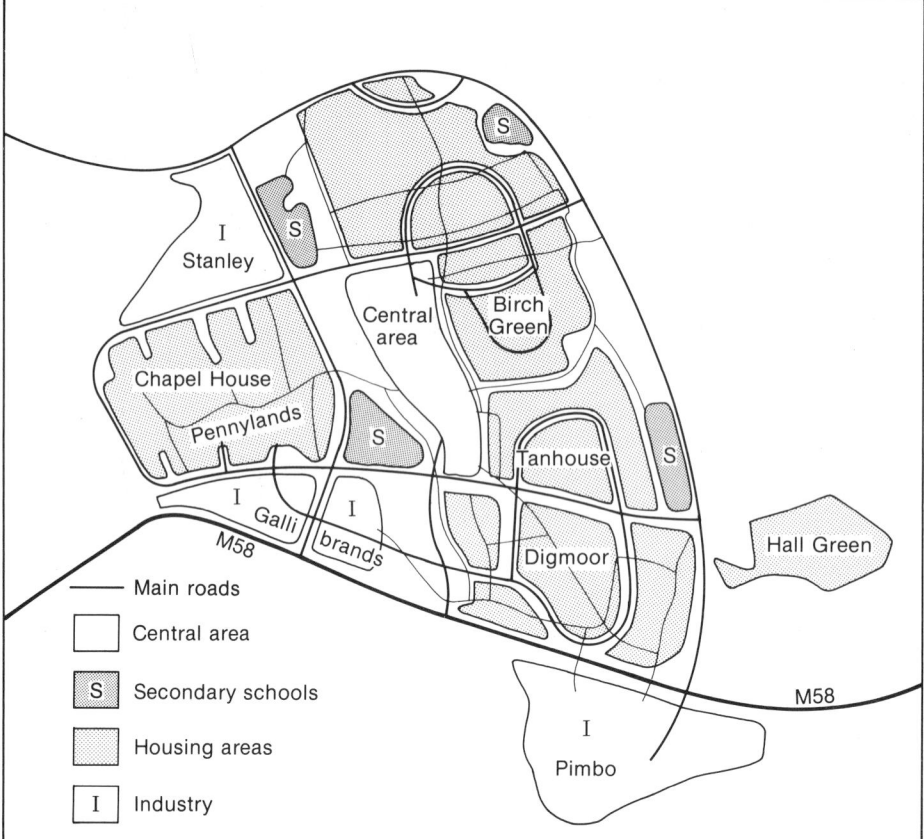

Figure 5.26. *Skelmersdale New Town*

Figure 5.27. *Skelmersdale concourse*
This is the central shopping area. It is totally enclosed and air-conditioned. Carefully located seating and plants provide a relaxed atmosphere

Population and Settlement · NEW TOWNS

Problems in Skelmersdale

There have been problems in Skelmersdale. Some families were not happy living in a new area, and returned to Liverpool. The houses they left empty were often damaged by vandals. For various reasons, some industry has closed down or moved to different areas. This has caused high unemployment, with little prospect of new jobs.

A further problem has been the high number of young people in Skelmersdale, which has put a strain on schools and health services. But there are not many old people. Those who live in the town may feel lonely and miss the old communities where they used to have their homes. Some people feel that New Towns are not very friendly places.

Figure 5.30. *Skelmersdale: the industrial site at Gillibrands*
Modern, clean factories are likely to attract firms to move to these sites. The main roads are fast, with several roundabouts linking them to other areas of Skelmersdale.

Figure 5.28. *Skelmersdale housing area of Ashurst*
A mixture of trees, shrubs and grass areas help make the areas more pleasant.

Figure 5.29. *Pedestrian route in Tanhouse, Skelmersdale*
There is a network of well-lit footpaths using subways and footbridges. This means that pedestrians can avoid crossing busy roads.

Questions

1 Describe the types of housing and their layouts in Figures 5.28 and 5.29.

2 Describe the layout, size and shape of the factories in Figure 5.30. What advantages do factory sites like this have when it comes to attracting new industry to Skelmersdale?

3 Much effort has been made by the planners to make Skelmersdale a pleasant place. Which features do you think people living in the town consider most important? What disadvantages might there be to living in Skelmersdale?

4 In the mid-1980s, there were no plans for further New Towns. Suggest reasons for this.

Regional Centres

In the hierarchy of settlements, regional centres lie below the capital and conurbations. As the name suggests, a regional centre is the main settlement in any large region of the British Isles. There are several cities and county towns within each region.

Figure 5.31 shows the regional centres in the southern part of Britain. Each town is fairly central to the region it serves. However, Cardiff is an exception. It is not in, or near, the middle of its region. It is convenient for people living in South Wales to use the city. Those living in the north of the country often find it easier to visit other centres, like Manchester or Birmingham.

These and other regional centres offer a wide variety of services and amenities. Many have local radio and television stations. Most support a morning and/or evening regional newspaper. Large department stores plus a good variety of smaller shops provide comprehensive shopping facilities. There are usually many office buildings. Big companies and banks often choose to have their regional, area, and local head offices in regional centres.

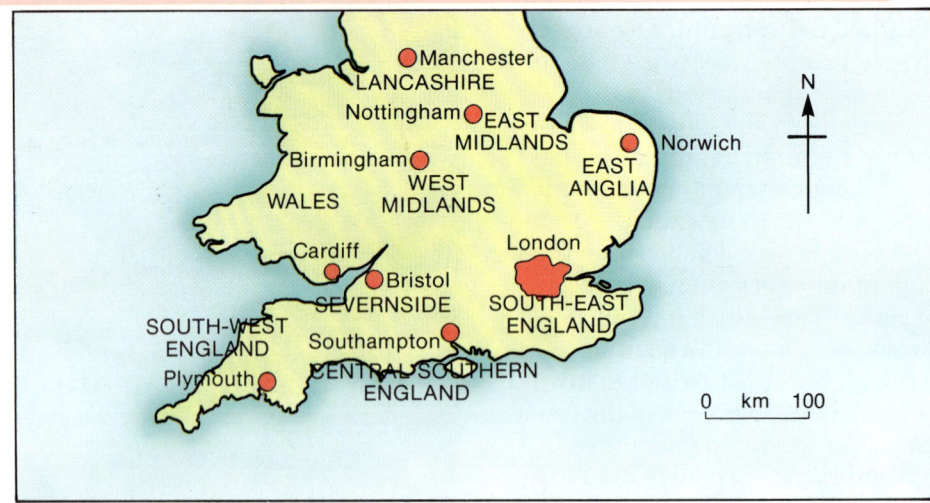

Figure 5.31. Regional areas in central and southern England, together with the areas they serve
Note: London is the capital and it also acts as a regional centre for south-east England

○ PLYMOUTH

Plymouth is the regional centre for south-west England. It is situated close to the border of Devon and Cornwall. The city has a long tradition as a naval port. It grew in size during the reign of Queen Elizabeth I. Plymouth's boat builders have provided the navy with many fighting ships. Famous sailors have been connected with the port. Drake sighted the Spanish Armada from Plymouth in 1588, and the first British settlers to America sailed the *Mayflower* from there.

The port of Plymouth is sited on a sheltered, deep water estuary. This has made it safe for ships to enter and leave the harbour. Until the 1950s it was a port of call for some of the world's largest ocean liners. However, Plymouth never became an important commercial port. Competition from Bristol and Southampton was too severe. These ports had better road and rail communications with the rest of Britain, and more local industry. At that time there was very little industry and few large centres of population in Devon and Cornwall.

During the Second World War, Plymouth was heavily bombed. The city centre was virtually destroyed. Post-war redevelopment has given Plymouth a much-admired new city centre. It has also provided a wide range of office accommodation. Such new facilities have done much to tempt new industry to the city from other parts of Britain.

Plymouth as a regional centre

Today, Plymouth is a regional centre. It is the largest city in Devon and Cornwall (see Figure 5.32). Both BBC and ITV have television studios in Plymouth. Their pictures are sent out to Devon and Cornwall, as well as parts of Dorset and Somerset. A regional newspaper, the *Western Morning News*, is also printed in Plymouth and covers a similar area. In addition, large department stores

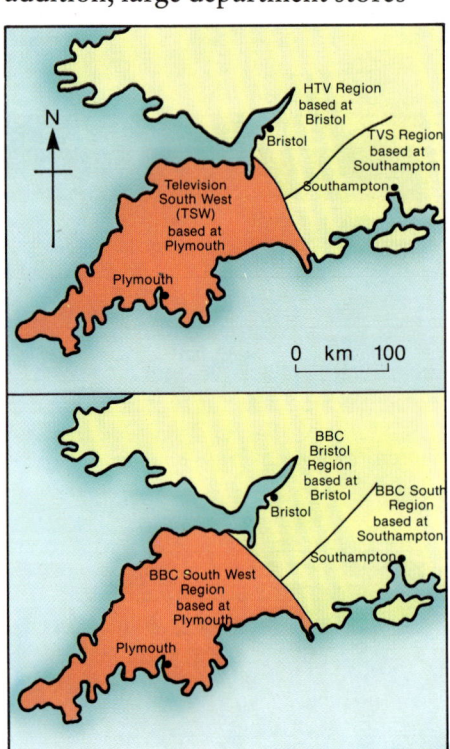

Figure 5.32. Plymouth – a regional centre
Maps (a)–(c) show some of Plymouth's different spheres of influence. Map (a) for example shows the area covered by independent television.
(a) Television South West (ITV)
(b) BBC South West

Population and Settlement · REGIONAL CENTRES

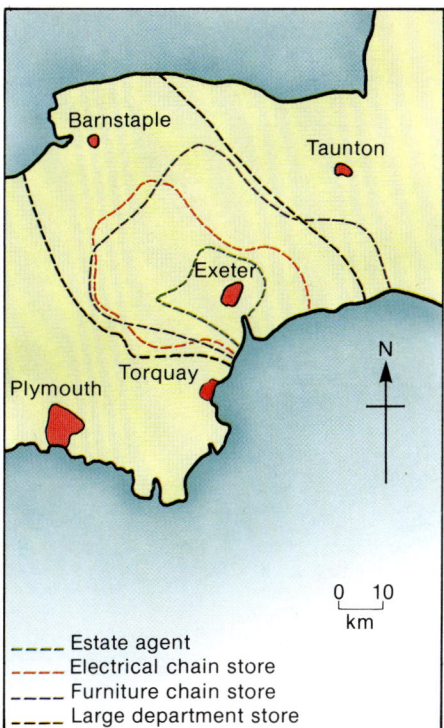

Figure 5.32. (c) Spheres of influence of some Exeter services

and other shopping facilities attract people from all over the south-west. Look at the maps in Figure 5.32.

The existence of Plymouth affects people in a wide area of south-west England. The fact that they make use of the city means that they come within Plymouth's 'sphere of influence'. The maps in Figure 5.32 show how far this sphere of influence extends. Beyond the boundaries, other centres provide similar services and amenities. Both Bristol and Southampton compete with Plymouth. You will notice from the maps that the sphere of influence for each item is slightly different. To some extent this shows the competition that takes place.

It is not only other regional centres that provide competition. Exeter has always been a very important centre. It originally had the only university in south-west England. Several large concerns have their head offices in Exeter, like the South-West Water Authority. Exeter is still the county town of Devon, complete with its own County Hall. The city also possesses many large department stores. Exeter therefore has its own sphere of influence, as the map shows. Dartmoor provides a natural barrier between the two settlements of Exeter and Plymouth. Many people prefer to use the facilities of Exeter rather than cross the upland area of Dartmoor to get to Plymouth.

Exeter can be considered as a sub-regional centre. Some of its services may actually be superior to those offered by Plymouth. As a result, the city is competing with Plymouth which has only recently established itself as the regional centre for the area.

Plymouth is still expanding. It is the largest city in Britain still increasing in population. All other larger and similar sized towns have declining populations. People are still being attracted to the city. One reason is that new industry offers job prospects. However, the south-west region is still the third worst area in Britain for unemployment.

○ GROWTH AREAS

In recent years there has been a movement of population from the north to the south in Britain. This has been caused partly by increasing unemployment in northern areas. In the south, industry tends to consist of smaller firms producing varied goods.

Within southern England there has been another movement of population from the London area to areas well outside the capital. Redevelopment of the inner city area has given people the opportunity to move away from London. Firms are moving to surrounding smaller towns too. Land is cheaper and there is room for expansion.

As a result of this, a belt of southern England has shown a marked increase in growth. Using your atlas, you should be able to follow it. Starting from Ipswich and Norwich in East Anglia, the growth area continues in a crescent through Cambridge, the new town of Milton Keynes, Oxford, Bath, and Swindon to the south coast around Southampton. Many of the towns in this area have increased in size.

This area is called 'the crescent'. It has the following features:
(a) It is between 80 to 180 km from the centre of London.
(b) There are improved roads and motorways to London (for example, the M4 from Swindon and the M1 from Milton Keynes).
(c) There are fast and regular rail routes into London (for example, 70 minutes from Southampton and 50 minutes from Milton Keynes).
(d) There are new housing areas.
(e) There is new industry, particularly high technology.

Questions

1 Study a map of Britain in your atlas. Name the regional centres in northern England, Scotland and Northern Ireland.
Regional centres
Manchester   London
Nottingham   Bristol
Birmingham   Southampton
Norwich   Cardiff

2 Copy out the table (below, left) which shows the main regional centres in southern Britain. Name the areas which they serve.

3 Why can London be considered a regional centre?

4 Describe the nearest regional centre to your school. Explain what makes it a regional centre and describe what facilities it offers.

5 Suggest reasons why Cardiff has become the regional centre for South Wales.

Holiday Resorts

With the coming of the railways, more people were given the opportunity of 'going to the seaside'. Trains enabled them to make day trips to the coasts. To begin with, most working people visited seaside towns like Scarborough and Blackpool closest to the urban centres. Blackpool was popular with people from Manchester, while Scarborough was favoured by day trippers from Leeds. As train travel became more popular, people started to go further afield. Many found that they could afford an annual holiday by the sea as well as a day out. Many seaside towns began to develop into the holiday resorts we know today.

The holiday makers needed somewhere to stay, so new hotels and guests houses were built in many coastal towns. As resorts attracted more people, they produced other amenities. Piers and theatres were early attractions. More recently, fun fairs and amusement arcades have been added. The larger resorts are shown on Figure 5.33.

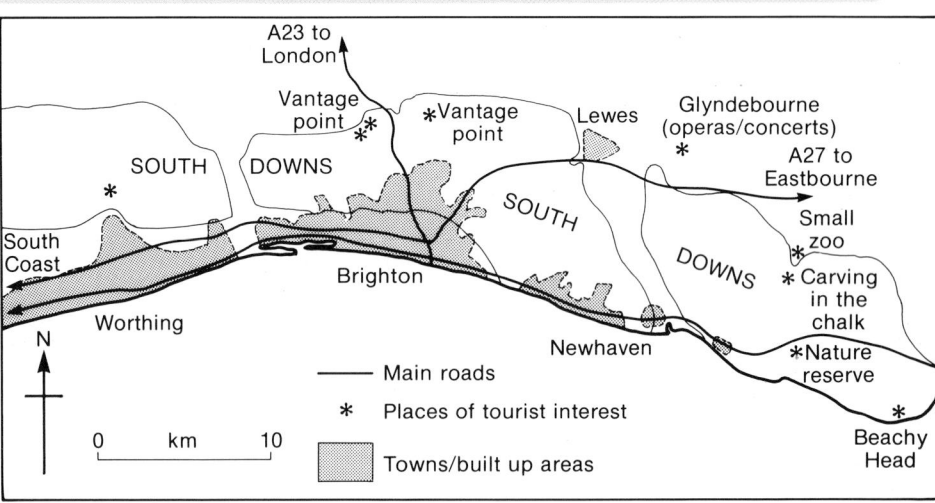

Figure 5.34. *The situation of Brighton, showing some places of interest to tourists*

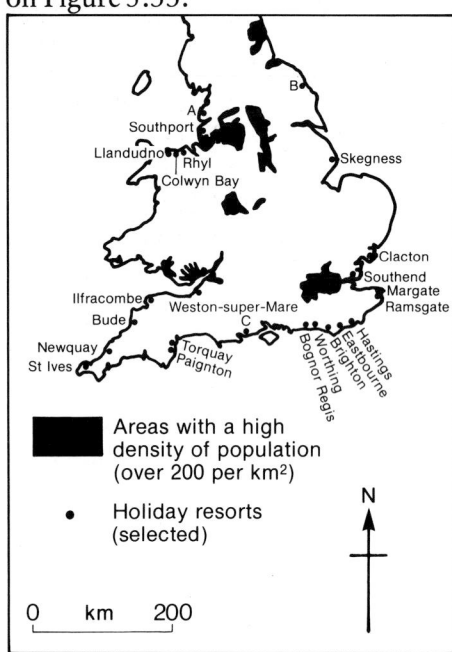

Figure 5.33. *Holiday resorts in England and Wales*
Using your atlas, name the three resorts A, B, and C

As transport networks have improved, people now have a very wide choice of resorts. Higher living standards mean that most people can afford to take a holiday. But not everyone wants to stay in a crowded resort. Many people want peace and quiet. Small fishing villages and coves around the coasts have also become popular, particularly in south-west England.

A large number of British holiday-makers go abroad each year. Staying in British resorts has become expensive. Often it does not cost much more to take a foreign package holiday. Sunny weather is more likely in the Mediterranean countries than in this country. Consequently, many holiday resorts in Britain have suffered a decline in their tourist trade.

A longer holiday season

Some tourist resorts have responded by trying to make their holiday season longer. This happens in the south of England, where it is slightly warmer than in the north. Large numbers of hotels now offer off-peak holidays in April, May, September and October. Many resorts have tried to attract new business, like conferences, during the rest of the year. This has reduced seasonal unemployment there. However, many hotels and businesses still close down when the beaches are deserted in the winter months. This puts people out of work.

All resorts need to have both 'dry weather' and 'wet weather' facilities. Dry weather facilities, like beaches, parks and golf courses are a great attraction when the sun is shining. Wet weather facilities, like cinemas, amusement arcades, sports centres and bingo halls, are important when the weather is cold and wet.

○ BRIGHTON

Along the south coast of Britain, there are numerous popular holiday resorts. Brighton is the largest and most important. Figure 5.33 shows the position of the others. Brighton is 82 km from London. Trains run hourly, taking 55 minutes, and the main A23 is a motorway part of the way.

Brighton was popular with the Prince Regent, later George IV, in the eighteenth century. He built a famous pavilion which was supposed to be the best in Europe.

Population and Settlement · HOLIDAY RESORTS   113

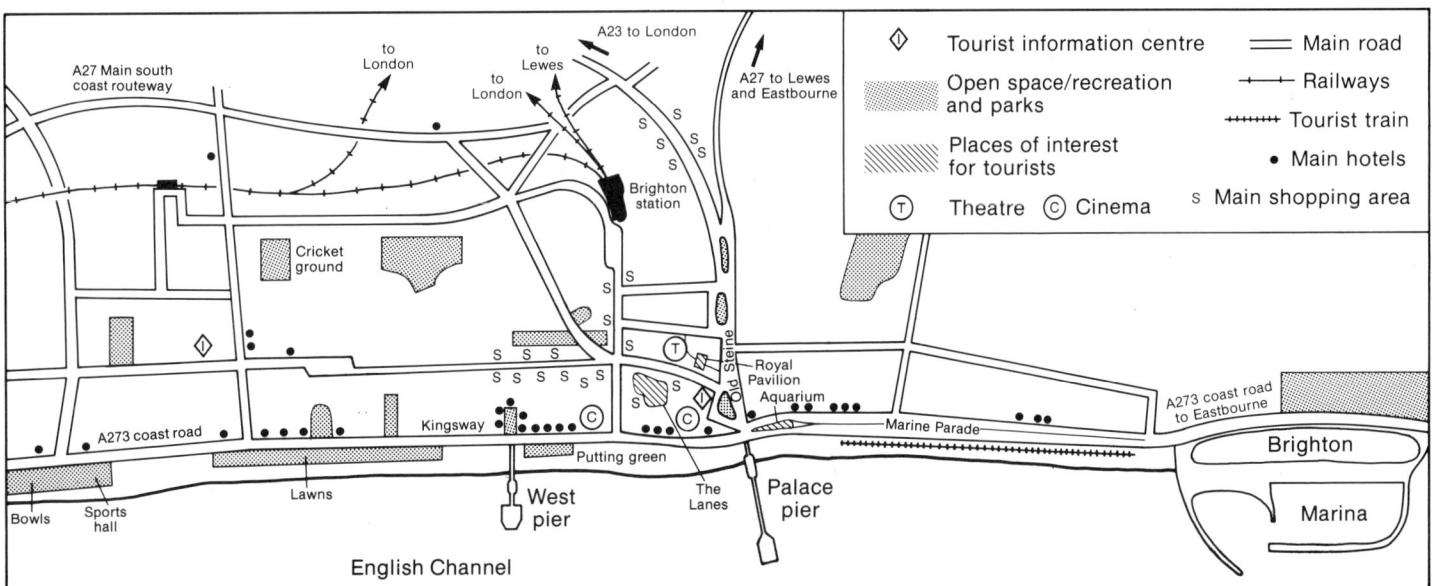

Figure 5.35. *Plan of Brighton*

Today, there is a large range of amenities, including the beaches; the small shops in the Lanes; the aquarium with dolphins as well as fish; and a variety of amusements along the sea front, including a pier.

One of the newest attractions is the yacht marina, which was completed in the late 1970s. It is one of the largest in Europe and attracts many people.

Around 7 million visitors a year come to Brighton. Of these, 6.5 million are day-trippers. Good parking facilities are needed. An efficient local transport network is necessary to take people to and from the various attractions. Visitors want to be able to get to the beaches when the sun shines, and the theatres, cinemas, and other indoor entertainments when it is wet.

Unemployment is a problem in many resorts, with large numbers of people out of work in winter months. But in Brighton, where many hotels are used as conference and meeting centres, the problem is less serious than it used to be. Figure 5.37 shows the increase in conferences and exhibitions held in Brighton. A large conference centre opened there in 1977.

Figure 5.36. *The coast at Brighton*

Questions

1 What advantages does Brighton have as a resort?

2 What is the difference between wet weather and dry weather facilities? Name four facilities of each type.

3 Name one resort that you have visited. What are its main attractions? List its wet weather and dry weather facilities.

4 How do you think Brighton might differ from a small Cornish holiday resort, in terms of:
(a) the number of daily visitors;
(b) where the visitors come from;
(c) the facilities offered?

5 Study Figure 5.37 and describe the trends shown by the figures. Did the opening of the new conference centre have much effect on these figures?

Figure 5.37. *Brighton as a venue for conferences*

Year	Number of exhibitions and conferences	Number of people attending
1970	185	63 000
1972	192	68 000
1974	246	90 000
1978	469	121 000
1980	1034	185 000
1982	915	138 000
1983	1041	127 000

Areas Where Few People Live

○ RURAL DEPOPULATION

Many people living in upland areas find life very difficult at times. A large number of negative factors deter people from moving to these areas. These difficulties include: relief (the high land and steep slopes); soils (which may be thin or acidic); vegetation (which may be heathland or moorland and useless for farming); communications (the lack of railway and road systems); and services (few shops, few hospitals or schools). There are fewer jobs in these areas. As a result most people prefer to live in other parts of the country.

There tends to be a drift of population away from isolated rural areas as people move to the towns in search of better job prospects. This is known as rural depopulation. Large areas of northern Scotland and central Wales experience depopulation.

Rural depopulation means that fewer people are left behind to work the land. Villages start to decline and become deserted. Young people tend to move away first so this means a high proportion of old people are left to look after themselves. Many become very isolated as more and more villages lose their shops, doctors, buses and other services that most of us take for granted.

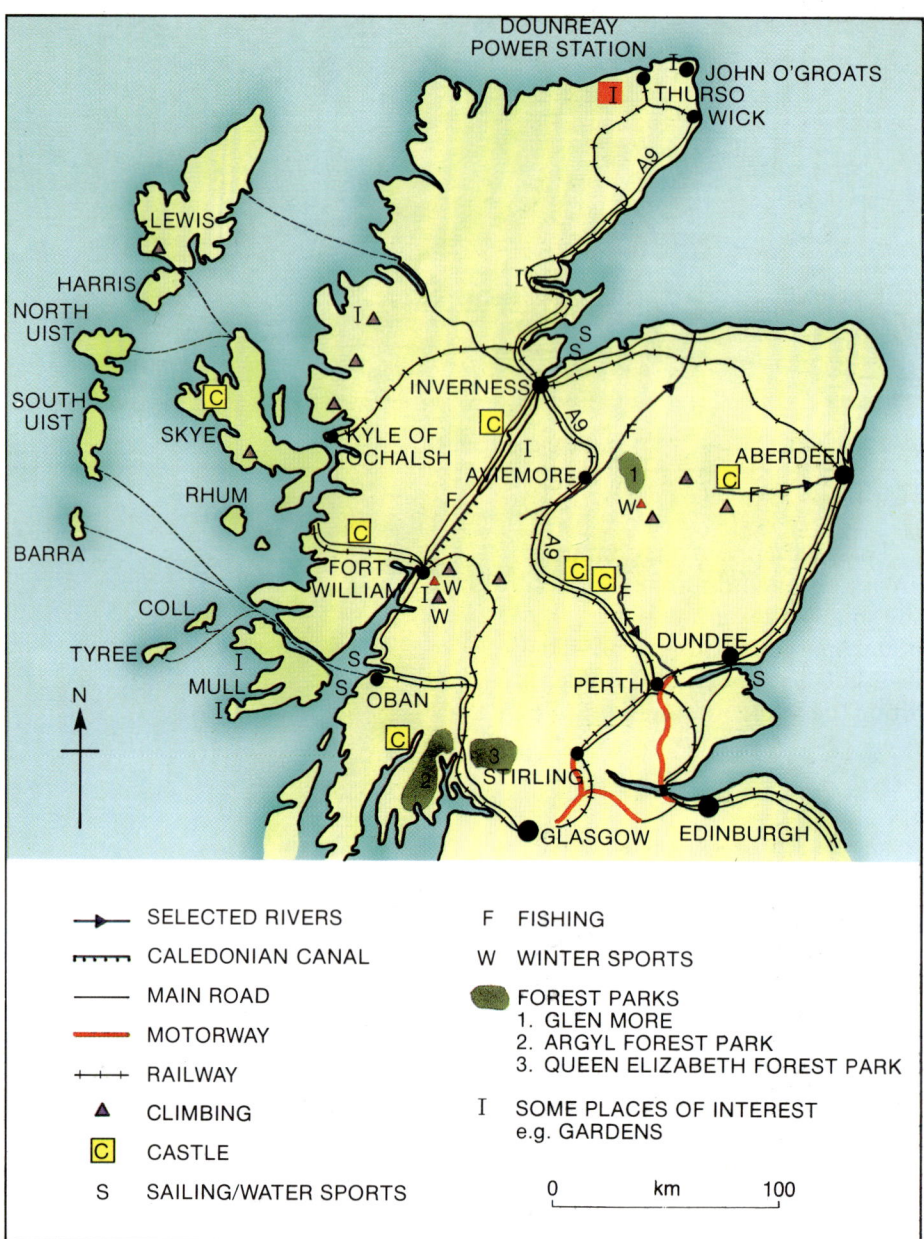

Figure 5.38. *Northern Scotland*

Exercise

Study the population map of Britain in an atlas. On an outline map of the country mark on the main areas with a population density of less than six people per km².

○ CENTRAL WALES

There has been a large movement of people away from central Wales. A New Town called Newtown has been built to try and stop the drift. New light industries have been introduced to provide more jobs.

As people have moved away from central Wales, others from England have bought the empty houses, cottages and farms to use as second homes or holiday homes. The properties are usually improved but they are only occupied for a short while each year. The buying of holiday homes pushes up the price of housing for the local community. This helps to force even more people to move away from the area, and encourages further depopulation.

Population and Settlement · AREAS WHERE FEW PEOPLE LIVE

NORTHERN HIGHLANDS AND ISLANDS OF SCOTLAND

In the north of Scotland, there are many negative factors which have caused people to move to other areas of Britain. Many parts of the region rise to over 1000 m. The land is often steep and the climate is harsh. Soils are thin and peaty, with much of the land covered in moorland vegetation. There are limited resources with few minerals. Only 15% of the land can be farmed.

In such poor conditions, a system of croft farming has been developed. A croft is a small piece of land about the size of a small field. Farmers, or crofters, mainly grow vegetables for their own use. They also raise some oats and hay for their livestock. Cattle provide milk and are occasionally used for meat. Sheep are kept mainly for their wool, although they sometimes supply the crofter with mutton. Some farmers also fish part-time to supplement their income. Today the fishing industry employs mainly full-time workers.

These harsh conditions forced many people to move away from the Highlands, to more prosperous areas in the south. Until recently 1000 people were leaving the area each year. Many of these were young people.

Figure 5.39. *Strathpeffer*
Once a railway station, this building now provides accommodation for small craft industries.

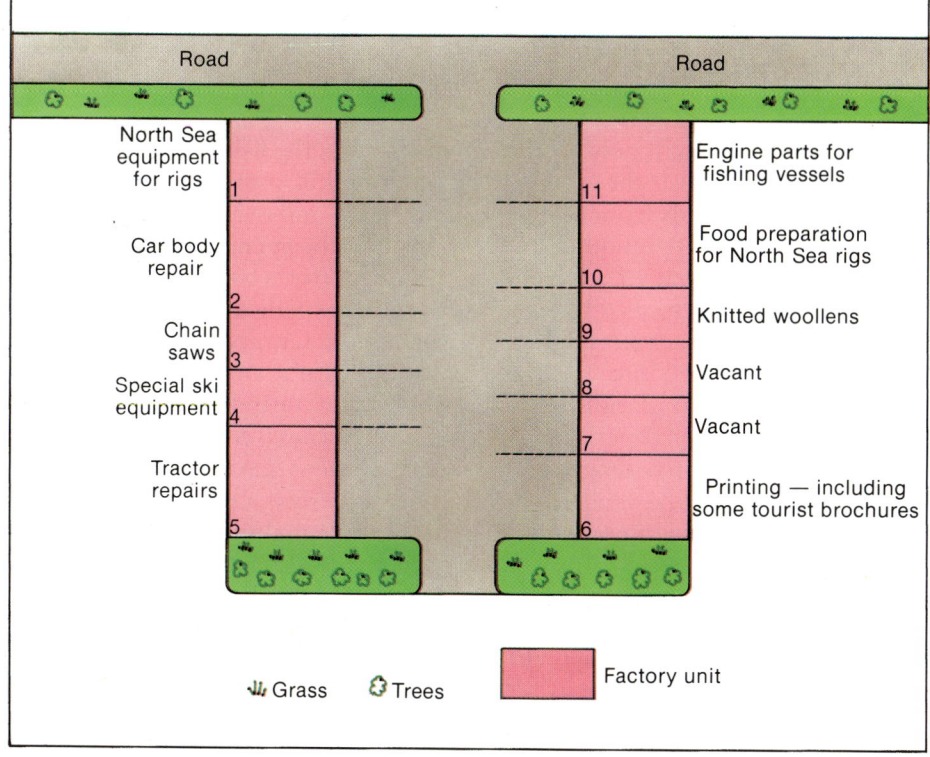

Figure 5.40. *A small modern industrial estate in the northern Highlands of Scotland, 10 km from the coast*
What do you notice about the size and shapes of the factory units? List those factories concerned with (a) North Sea oil and fishing; (b) farming and forestry, and (c) tourism. Suggest two different industries that might use factory units 7 and 8. From the map, state ways in which the estate is made pleasing in appearance.

Figure 5.41. *The Stornaway ferry, which provides a very important link with the Islands*

Figure 5.42. *Light aircraft*
Small aircraft provide a fast service to places like Glasgow, but they can only carry a few passengers and a limited number of goods.

Improvements in the Highlands and Islands

To combat this population drift, the Highlands and Islands Development Board was set up. Its aim was to try and improve the way of life in this part of Scotland. It planned to do this by attracting industry; improving farming; attracting tourism; providing facilities for new craft industries and by developing an efficient transport system.

Today, new factory estates have been built around towns to try and reduce unemployment, (as Figure 5.40 shows). The industrial estates all lie along main roads and near to towns.

Tourism has brought new wealth to the area. People visit the Highlands and Islands from many countries, as well as from other parts of Britain. About 15% of all visitors are from overseas. Hotels have been improved, chalets built and old houses turned into self-catering accommodation. This helps to bring some prosperity to the area, as well as providing valuable jobs. Skiing has helped to make this a popular winter sports resort.

Fishing has also been given financial aid. New boats are very expensive, so grants have been given to help younger people to fish. In 1978, over £4 million was invested in the fishing industry. As a result of this, more boats now put out to sea and over 200 full-time jobs have been created.

Great efforts have been made to introduce small craft industries to the region. Grants have been made available for many small firms. Highland Craftpoint, about 16 km from Inverness, is a new development service for the crafts industry. Glass, ceramics, pottery and candle making are some of the newer crafts encouraged in this area.

The Highlands and Islands Development Board has encouraged crofters to join together to form small groups. These

Population and Settlement · AREAS WHERE FEW PEOPLE LIVE

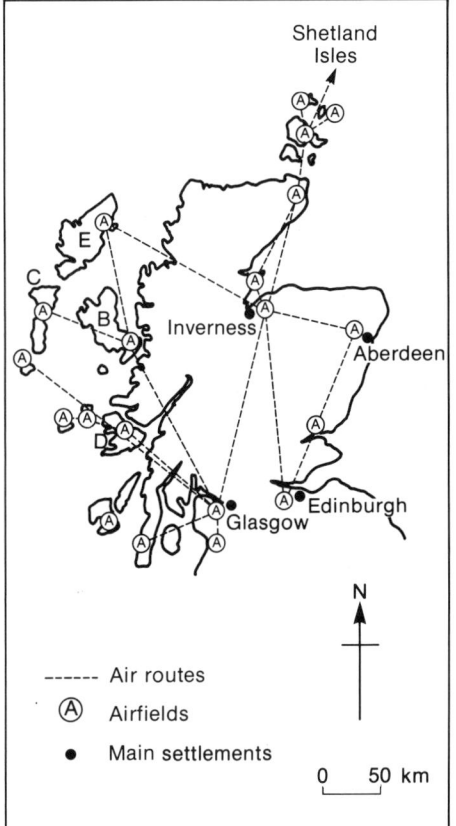

Figure 5.43. *Domestic air routes in Northern Scotland*
Using your atlas, name the islands B, C, D and E. What is the distance in km from island E to Glasgow? If there were no air service, describe the problems involved in making a journey from island E to Glasgow.

Figure 5.44. *Winter service of ferries (Reproduced by permission of Caledonian McBrayne)*

BARRA and SOUTH UIST/OBAN * CASTLEBAY * LOCHBOISDALE

OBAN .... dep.	0030	0030	LOCHBOISDALE	0730	1015
CASTLEBAY ..... arr.	0930L	0630	CASTLEBAY	0930	—
CASTLEBAY ..... dep.	—	0715	CASTLEBAY	1015	0715L
LOCHBOISDALE .. arr.	0630	0915	OBAN	1645	1645

SMALL ISLES/MALLAIG * EIGG * MUCK * RHUM * CANNA

	Monday	Wednesday	Thursday & Saturday
MALLAIG	1230	1230	0500
ARMADALE	1300	1300	1130
EIGG	1430	1430	0945
MUCK	—	1515	—
RHUM	1600	1630	0830
CANNA	1700	1730	0730
ARMADALE	—	—	—
MALLAIG	1930	2000	1200

NORTH UIST AND HARRIS/LOCHMADDY (N. Uist) – TARBERT (Harris) – UIG (Skye)

		Monday	Tuesday Thursday	Wednesday Friday	Saturday		
LOCHMADDY	dep.	—	1400	0630	—	—	1330	0630	1130
TARBERT	dep.	0630	—	—	1330	0630	—	—	1400
UIG	arr.	0830	1600	0845	1545	0845	1545	0830	1600
UIG	dep.	0900	1615	0930	1615	0930	1615	0900	
TARBERT	arr.	1100	—	1200	1830	—	—	1330L	
LOCHMADDY	arr.	1330	1830	—	—	1200	1830	1100	

LEWIS/ULLAPOOL – STORNOWAY

	Mon. & Sat.	Wed. & Fri.	Tues. & Thurs.	
ULLAPOOL ..... dep.	1200	0930	0930	1730
STORNOWAY ... arr.	1530	1300	1300	2100
STORNOWAY ... dep.	0800	0530	0530	1330
ULLAPOOL ..... arr.	1130	0900	0900	1700

operate like co-operatives with crofters sharing equipment and machinery. This helps them to cut their costs. Although only on a small scale at the moment, there are possibilities that such co-operatives will expand. Spinning and weaving wool into local cloth could be one idea.

Forestry has been encouraged. Large areas of once barren moorland have been converted to young coniferous forest.

Finally, the Board has made great efforts to improve communications. Roads have been widened and straightened, particularly the important A9. Air services have been improved, and so have ferries to the islands. Few improvements have been made to rail services, but large sums of money have been provided to keep trains running on social grounds.

Today, the population in this area is fairly constant. One very important reason for this is North Sea oil. It has attracted workers to the oil rigs, and also to the many related service industries.

Questions

1 Draw a sketch map of the west coast of Scotland from Oban northwards, and include the main islands. Name the islands and mark on the ferry routes given in the tables above. Name the towns at the end of each route, and add the time taken for each ferry journey.

2 What do the symbols at the top of each table mean?

3 Describe the goods that might be transported on these ferries.

4 What are the problems of having to rely on a ferry service?

National Parks

Certain areas of England and Wales have been chosen to become National Parks. They were selected because they are areas of great natural beauty. One area in Wales that attracts large numbers of tourists is Snowdonia National Park. It is an area with a sparse population, where special care has been taken to conserve the natural beauty.

The main aims of all national parks are:
(a) the preservation and advancement of natural beauty;
(b) the promotion of their enjoyment by the public.

The landscape in Snowdonia is rugged and mountainous, with U-shaped valleys carved by ice. Some of Wales' most important peaks are found here, like Snowdon and Cader Idris.

Figure 5.46. *Walking over Cwm Bochlwyd in Snowdonia*
Walkers of all ages are attracted to areas like this. Suggest reasons for this attraction. Why is the northern part of Snowdonia more popular than the southern part?

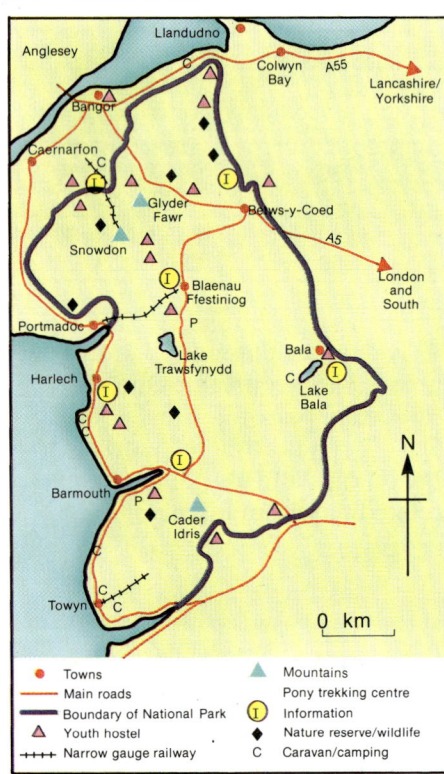

Figure 5.45. *Snowdonia National Park Using your atlas, describe the various ways in which Snowdonia National Park may be reached. What problems will a large number of tourists cause on the transport routes? How might the problems be overcome?*

The bleak and rugged mountains have not attracted many people to live in this part of Wales. Those who have settled often find it very difficult to make a living from the poor farmland high on the hillsides. Some farmsteads on the upper slopes have been empty for many years. So, as in the Highlands of Scotland, rural depopulation has been a problem in Snowdonia.

○ IMPORTANT INDUSTRIES IN SNOWDONIA

Industry in the area has changed. A hundred years ago slate quarrying was very important. The slate was used mainly for roofing and was sold to customers in North America as well as in Britain. There were also many small gold, lead and copper mines. Some had been worked for 300 years. Most of these industries built small railways to transport the minerals and slate to the coast for shipping. The remains of many of these lines can still be seen. Mining and quarrying have declined, although some slate mines and railways have been re-opened as tourist attractions. Tourism is now one of the main industries in Snowdonia.

Farming is important as well. Conditions are difficult, but there is grazing for sheep in upland areas during the summer. The wide, sheltered valleys provide pasture and arable land, and some barley is grown in western valleys. Hay is harvested in the Conway valley.

Forestry is the third main industry in Snowdonia. Once the whole area was covered by trees. Later, large numbers were cleared by people seeking land on which to graze sheep. However, many trees are being replanted, with one tenth of the National Park now covered by forest.

Tourists can travel to the National Park by road or rail. Too many cars in the Park would cause problems. A Snowdon Sherpa Bus Service is provided, to encourage motorists to 'park and ride'. Re-opened narrow railway lines also transport visitors.

Population and Settlement · NATIONAL PARKS

○ OTHER NATIONAL PARKS

There are ten National Parks in England and Wales. Seven are in England, and three in Wales (see Figure 5.47). They were set up in the 1950s, in an attempt to preserve the natural beauty of the countryside. They were created in areas which could be reached easily by people living in neighbouring towns and cities. For example, the Peak District can be reached by people in Manchester and Sheffield. Motorways and general road improvements have made travel much easier and quicker. London is the only city which is a considerable distance from any National Park. People in Manchester can reach any one of five Parks within a day.

The better road communications have created problems. Some areas of the Parks become very congested in the holiday season. This is one reason why the bus services, mentioned earlier, have been introduced. The buses link the more popular parts of the parkland and enable visitors to leave their cars around the edge of the Park. This relieves the congestion inside.

Great efforts have been made to provide amenities in National Parks, without destroying the natural landscape. Such facilities include car parks, camping sites, cafés and souvenir shops. However, it is often difficult to make the city-dwellers aware of the countryside. They often trample over crops, leave gates open, and let their dogs worry sheep, all of which annoy local farmers. There is a Country Code which should always be followed when visiting the countryside.

The National Park is also responsible for preventing the area from being spoilt. It controls the use of the land and the types of buildings erected. Care needs to be taken when new buildings are planned so that the character of the landscape is not changed. For instance, a large brick and tile house would be out of character in a village built of stone and slate.

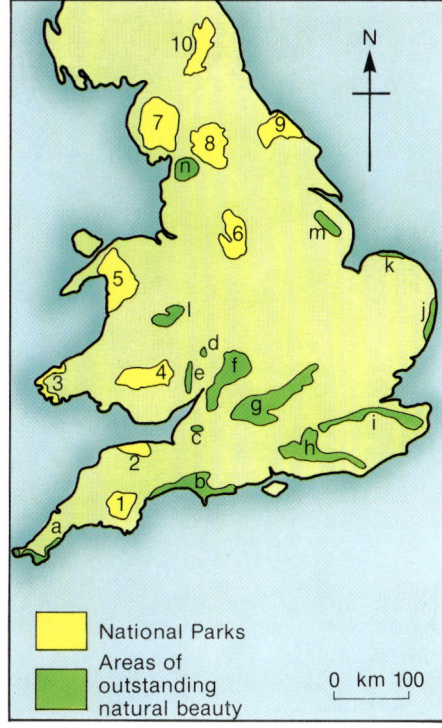

Figure 5.47. *The National Parks*

Occasionally, there is conflict between industry and the Park. An excellent example exists in the Peak District, where limestone is quarried. The amount taken from the ground has doubled every twelve years since the War.

The National Parks

The 10 National Parks are:

Brecon Beacons	1344 km²
Dartmoor	945 km²
Exmoor	686 km²
Lake District	2243 km²
Northumberland	1031 km²
North Yorkshire Moors	1432 km²
Peak District	1404 km²
Pembrokeshire Coast	583 km²
Snowdonia	2174 km²
Yorkshire Dales	1761 km²

The main areas of outstanding natural beauty are:

Lincolnshire Wolds
Norfolk Coast
Suffolk Coast
Chilterns
Cotswolds
Surrey Hills and Kent Downs
Sussex Downs
Dorset
Cornwall
Mendip Hills
Malvern Hills
Wye Valley
Shropshire Hills
Forest of Bowland

Questions

1 Study the map of Snowdonia and the photographs. Describe the facilities offered to tourists in this area.

2 On an outline map of England and Wales, copy the areas marked on Figure 5.47. Shade National Parks in red, and areas of outstanding natural beauty in green. Using an atlas, name the lettered and numbered areas. The names are all in the above list.

3 Mark on the main cities of England and Wales. What do you notice about the position of most of these towns in relation to the National Parks? Are any cities a long way from a National Park?

4 What are areas of outstanding natural beauty? Find out more about the one which is nearest to your school. Try to explain why it was set up, why it is an area of outstanding natural beauty and what facilities are available to tourists.

5 Imagine that you were allowed to choose an area of outstanding natural beauty to become a National Park. Which area would you choose? Give reasons for your choice. You might include such factors as: nearness to a large town; easy access by public transport or private car; attractive scenery; other facilities provided, such as parking or refreshments.

Round Britain Race ....

You will each need a die and a pointer. It is best to work in small groups. Each player throws a die in turn and moves along the road. Each shaded square is a chance square. If you land on a chance square read the instructions by the side of the map. Follow the route on an atlas and on each round describe the towns that you have passed through. Also, describe any chance events.

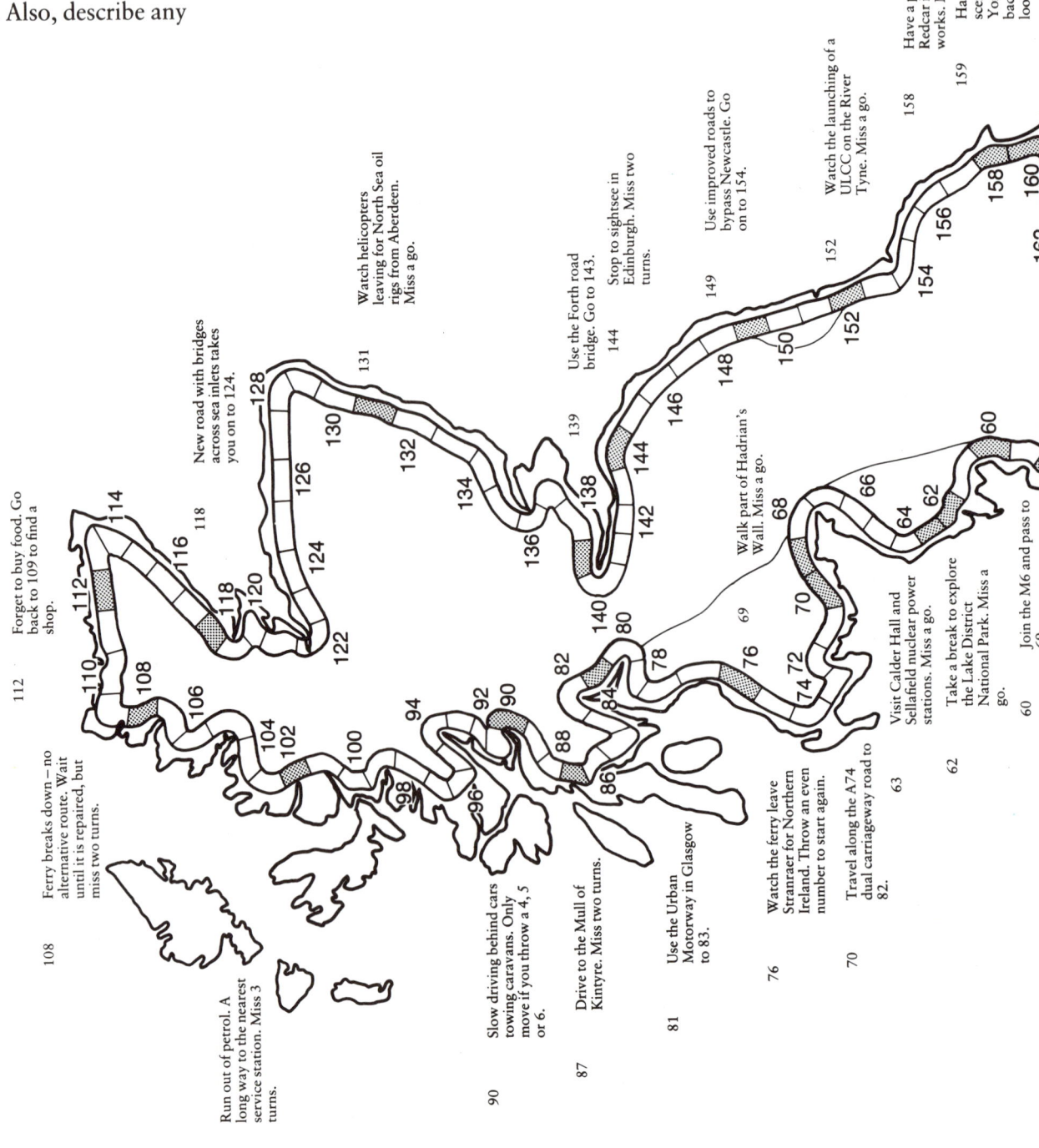

ROUND BRITAIN RACE

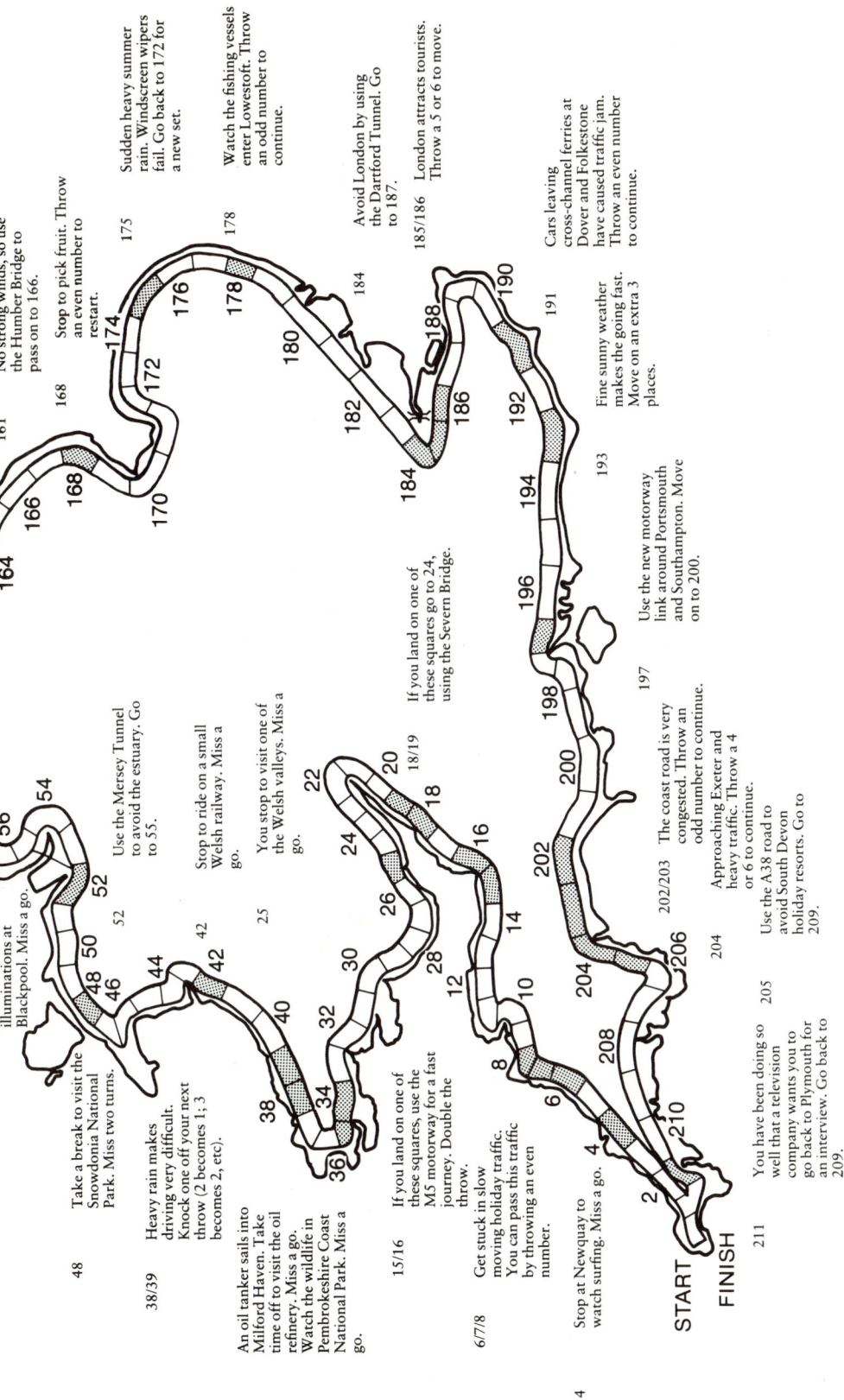

Map Reading Questions

1 Ellesmere Port map (Map 1) (Sheet 117 1:50 000 Landranger Series)
  (a) Give the six figure grid reference of:
    (i) Stanlow Railway Station;
    (ii) the Church with a tower at Thornton-le-Moors.
  (b) What can be found at the following grid references:
    (i) 394736
    (ii) 405746
    (iii) 400759
    (iv) 398777
  (c) State two reasons why there is no settlement in grid square 4373.
  (d) What is the direction of Ince (grid square 4576) from Stoak (grid square 4273)?
  (e) Suggest reasons why a sewage works is found at 424743.
  (f) Describe the drainage in grid square 4373.
  (g) Locate the following grid references: 448729, 398754, 427727. Pair each grid reference with one of the following descriptions:
    (i) a hall completely surrounded by housing;
    (ii) a large house 15 metres above sea level and near to a C class road;
    (iii) a farm close to a B class road and a stream on the opposite side of the road to another farm.
  (h) Imagine you are staying on a Geography Field Course in this area. You have three full days in the area. Suggest a course that you would undertake, stating your aims, where you would go, what you would do and the conclusions you might expect to reach.

2 Cairngorm map (Map 2) (Sheet 36 1:50 000 Second Series)
  (a) Give the six figure grid reference of
    (i) the public telephone at Coylumbridge
    (ii) the car park on the shore of Loch Morlich.
  (b) What can be found at the following grid references:
    (i) 977098
    (ii) 958058
    (iii) 927078
    (iv) 988062
  (c) Why does the road from 985080 to 987070 take such a long route?
  (d) How does the River Luineag and its valley in grid square 9410 differ from the Allt Druidh and its valley in grid square 9406?
  (e) State two reasons why there is little settlement in grid square 9305.
  (f) Give two reasons why there are seasonal ski tows and chair lifts in the south-east part of the map.
  (g) Locate the following grid references: 976097, 970078, 952068. Pair each grid reference with one of the following descriptions:
    (i) a lodge on the edge of the forest at the base of a very steep slope;
    (ii) caravan park in the forest;
    (iii) track with a stream beside it in the forest.
  (h) Imagine you were staying at Glenmore Lodge (9809) on a Geography Field Course. You are going to spend one day studying one stream or river, and another day studying the tourist facilities. For each day, state what you would do, where you would do it and the methods that you would use to undertake your studies.

3 Rugeley map (Map 3) (Sheet 128 1:50 000 Landranger Series)
  (a) Give the six figure grid reference of
    (i) the church with a spire in Brereton
    (ii) the church with a tower in Armitage.
  (b) What can be found at the following grid references:
    (i) 071175
    (ii) 025159
    (iii) 086144
    (iv) 039157
  (c) Give two reasons why the site at 043128 was chosen in ancient times as a fort.
  (d) Compare the route of the railway between Rugeley and Handsacre with that of the A513 road.
  (e) State TWO reasons why a settlement developed at Handsacre.
  (f) What evidence is there to suggest that Rugeley is a route centre?
  (g) Locate the following grid references: 047177, 042168, 057168. Pair each of the grid references with one of the following descriptions:
    (i) an industrial site;
    (ii) an area of old housing;
    (iii) a new housing estate.
  (h) Imagine you are to spend two days studying the town of Rugeley. State the different aspects that you would study; how you would do it, and how you would present your findings.

Map Reading Questions

Map 1: Ellesmere Port..................

Map 2: Cairngorm

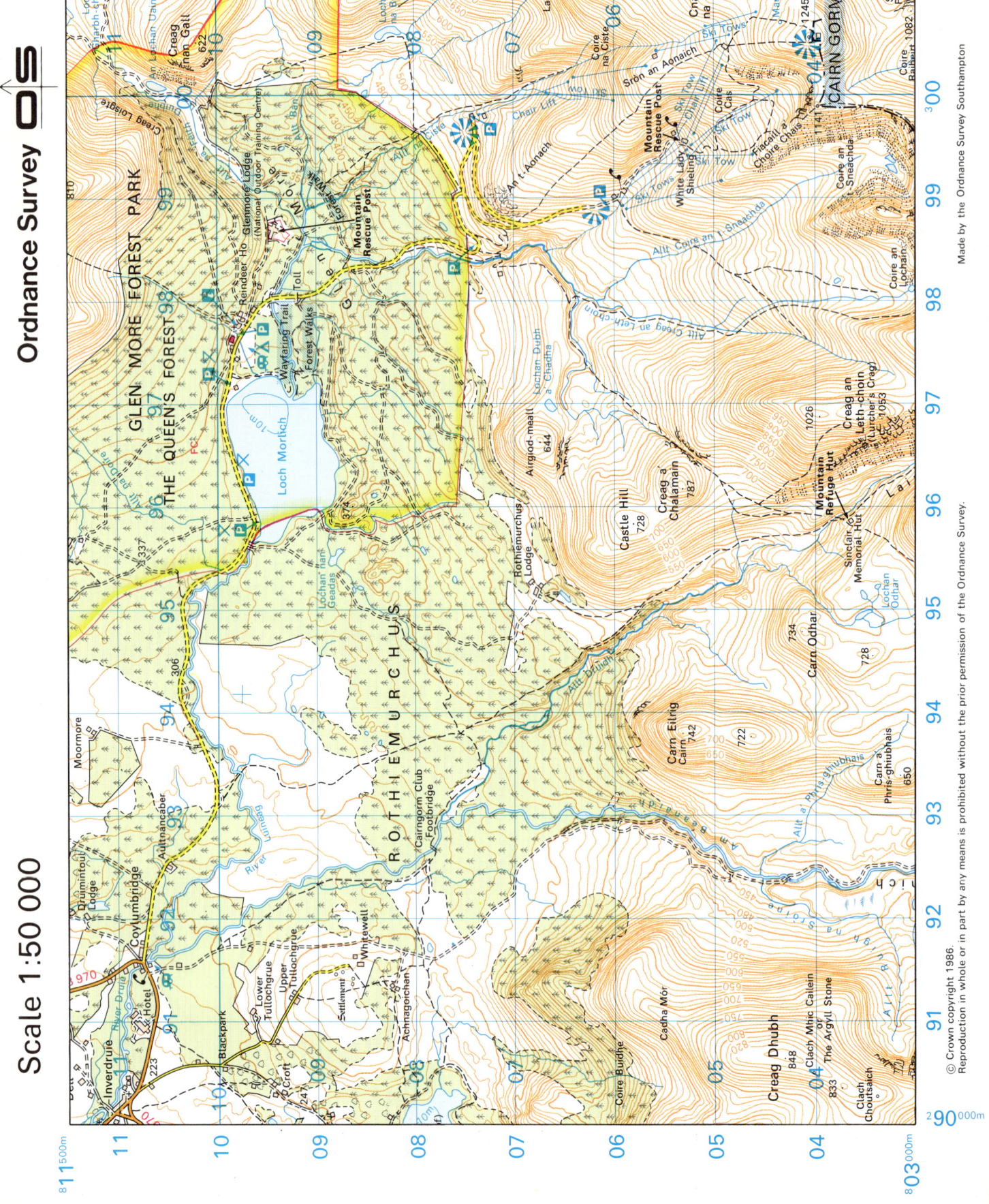

Map Reading Questions

Map 3: Rugeley

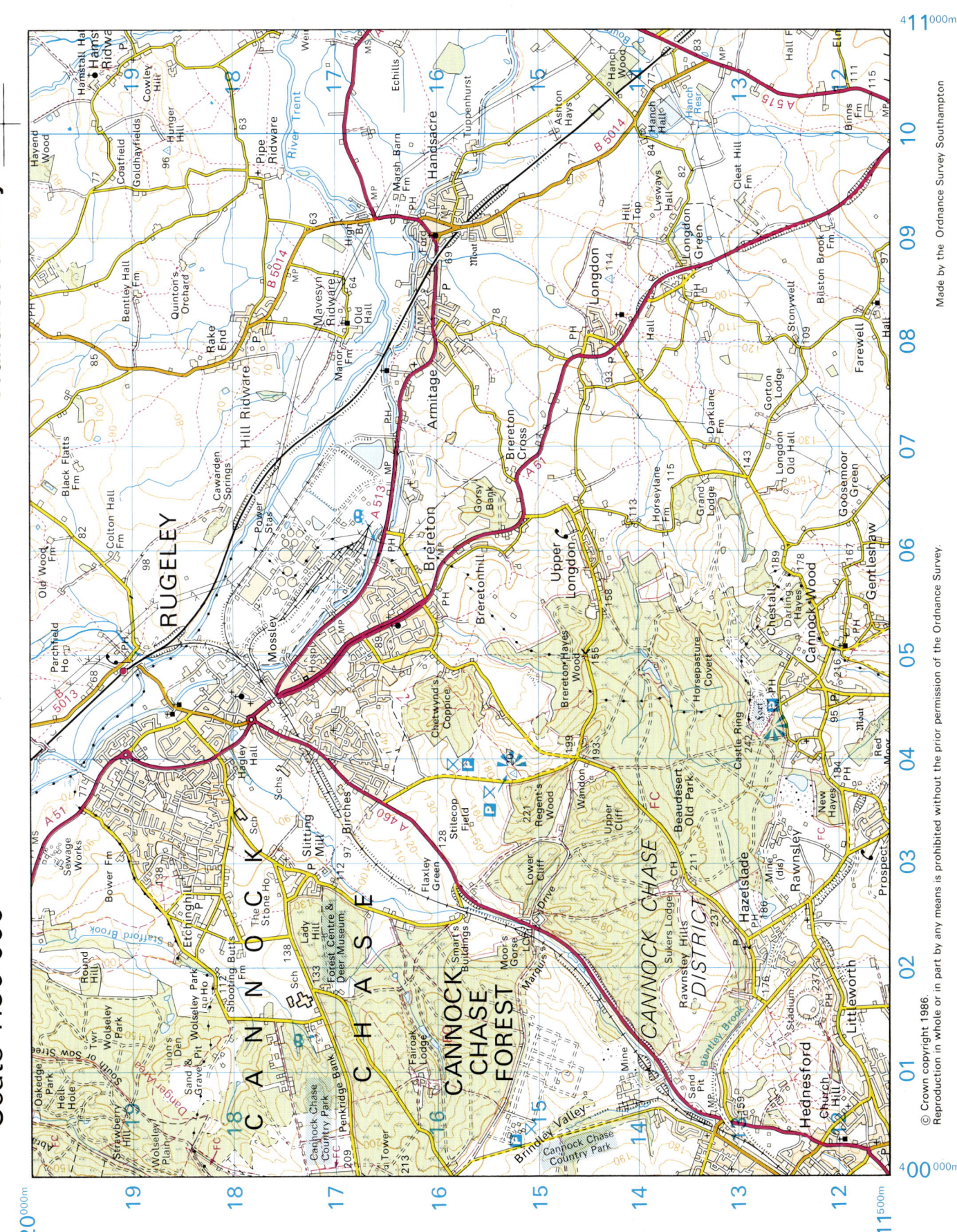

Index

Numbers in bold refer to illustrations

airports, 92–3, **92**
aluminium, 55, **55**
aquifer, **16**, 17
arable farming, 36, 42–5, **42**, **43**, **44**, **45**

blast furnace, 51–2, **52**, **53**
Brighton, 112–13, **112**, **113**

Cannock Forest, 32, 34–5, **34**, **35**
CAP, 39
car industry, 62–5, **62**, **63**, **64**, **65**, 76, 78
cement, 14–15, **14**, **15**
central business district, **100**, 101
Central Wales, 17, 85, 114–15
cereals, 36
chalk, 13–15, **13**
chemicals, 58–9, **58**
Cheshire, 40–1, 58
cities, 98–9, **98**, **103**
Cleveland, 53, **59**, 60–1
climate, 42, 46, 48
coal, 1, 2–7, **2**, **3**, **5**, 20, 26, 58, 61, 68
  bituminous, 2
  coking, 2
coalfields, 2, 4, **4**
  concealed, 2, **2**
  exposed, 2, **2**
  Selby, 6, 86
  South Wales, 4, 7, 74
  Yorkshire, 5
coal fired power stations, 25, **25**
coke, 51
Common Agricultural Policy (CAP), 39
conservation, 33
containers
  by sea, 89, **89**, 91
  by train, 86, **87**
conurbations, 102–3, **102**
Cornwall, 18–19, **18**, **19**
cotton, 68–9, **68**, **69**, 76
Crawley, 106–7, **106**, **107**

dairy farming, 38–41, **38**, **39**, **40**, **41**
depopulation, 114–15
development areas, 74–5, **74**
Dungeness, 24, **24**

East Anglia, 44, 66
electricity, 20–7, **20**
energy, 20–7, **26**
Enterprise Zones, 74, 75

farming, 36–49, **37**
  see also arable; dairy; hill
fish farming, 31, **31**
fishing, 28–31, **28**, **29**, **30**
forestry, 32–5, **32**, **33**, 117

gas, 8, **9**, 20, 26, **26**, 30, 117
Gatwick, 92–3, **93**, **94**, 106
geothermal power, 27
government, 48, 70–1, 74–5
Green Belt, 106
growing season, length of, 36

hamlet, **98**, 99
hardwood, 32–3
Heathrow Airport, 92–5, **94**, **95**
HEP see hydroelectric power
hill farming, 46–7, **46**, **47**
holiday resorts, 112–13, **112**
Hydroelectric power (HEP), 20–2, 21, 26, 55

industry, 50–77, **73**
iron and steel, 50–5, **52**, **53**, **54**, **55**
iron ore, 50–1, **50**, **51**, 59

Lancashire, 68–9
light industry, 68, 71, 72–3, **72**, 97, 108–9, 115–16
limestone, 12–13, **12**, **13**
Loch Awe HEP scheme, 21–2, **22**
London, 80–1, 88–9, 106, 111
  Port of, 90, **90**
Longbridge, 64–5, **64**, **65**

market gardening, 36, 37, 48–9, **48**, **49**
mining, 2–3
motorways, 64, 79, 80–1, **80**, **81**, 91, 102, 111

National Parks, 118–19, **119**
New Towns, 101, 106–9, **106**, **107**, 114
North Midlands, 100–1
North Nottinghamshire coalfield, 5, 6
North Sea oil and gas, 8–11, **9**, 30, 117
North/South divide, 76–7, **76**
North West England, 17
Northern Highlands, 115–17, **115**, **116**, **117**
nuclear power, 20, 22–4, **23**, 26

oil, 8–11, **8**, **9**, 20, 58, 60, 69
  refining, 56–7, **56**, **57**, 61
oil seed rape, 43
open cast mining, **3**, 51

Peak District National Park, 13, 47, 119
petrochemicals, 60–1
pipelines, 57
plutonium, 22–3
Plymouth, 110–11
population, 96–7, **96**, **97**, **104**, **105**, 111
ports, 88–91, **88**, 110

railways, 61, 82–7, **82**, **83**, **84**, **85**, **86**, **87**, 114, 117
  freight trains, 86, **86**
  passenger services, 82–3, **84**, **85**
  uneconomical lines, 85, **85**
recreation, 33, 35
Redcar, 53–4
regional centre, 110–11, **110**
reserves, 1
resources, 1
roads, 61, 78–81, **78**, **79**, **80**, 101, 114, 117
rural depopulation, 114–15

Scottish Highlands, 21–2, 115–17
Selby coalfield, 6–7, **7**, 86
settlement, 98–9, **99**
sheep farming, 36
shipbuilding, 54, 70–1, **70**, **71**, 76
Skelmersdale New Town, 108–9, **108**, **109**
Snowdonia, 118–19, **118**
softwood trees, 32–5
soils, 36, 38, 42, 46, 48, 114
solar power, 27
South East England, 97
Stanlow oil refinery, 56–7
steelworks, 50, 52–5, 74, 76
Sullom Voe, 10–11, **11**
synthetic fibres, 66, 68–9

Teesside, 60–1, **60**
textiles, 66–9, **66**, **67**, **68**, **69**
Thames Basin, 17
tidal power see wave power
Tilbury, 90–1, **91**
timber, uses of, 33
tourism, 35, 112–13, 116, 119
towns
  functions of, 100–1
  inside, 101
  see also New Towns
transport, 78–95
Tyne River, 70–1, **71**, 76

ULCCs, 89, **89**
unemployment, 74–5, **75**, 77, 109, 116
uranium, 22–3

water, 16–19, **16**, **17**, **19**, 96, 111
wave power, 26, **26**
West Midlands, 62–5, 73, 84, 102–4
woollens, 66–7, **66**, **67**, 117

Yorkshire, 66–7, **66**, 69, 97